THE NEW RIVER
CONTROVERSY

Thomas J. Schoenbaum

THE NEW RIVER CONTROVERSY

With a foreword by former senator
SAM J. ERVIN, Jr.

JOHN F. BLAIR, PUBLISHER
Winston-Salem, North Carolina

Library of Congress Cataloging in Publication Data

Schoenbaum, Thomas J
    The new river controversy.

    Bibliography: p.
    Includes index.
    1. Water resources development—New River Valley,
N.C.-W. Va. I. Title.
  HD1695.N53S36   333.9'1'00975623   79–24108
  ISBN 0–89587–008–8

43,322

To JEFF and LIZ

Foreword

THOMAS J. SCHOENBAUM'S *The New River Controversy* is most entrancing. Although it was written for the primary purpose of describing the conflict between energy development and environmental quality occasioned by the effort of the Appalachian Power Company to dam the New River for hydroelectrical purposes, the book portrays the geologic forces which created the New River, the prehistoric coming of the Indian to the area, the history of the American Revolution relevant to it, and the way of life of the people of Ashe and Alleghany counties in North Carolina, and Grayson County, Virginia, for a space of two centuries.

I absorbed an undying love for the New River area from my loyal friend, Floyd Crouse, a native of Alleghany County, who attended the University of North Carolina at Chapel Hill and Harvard Law School with me for six years. After being licensed to practice law, Floyd returned to his home in Sparta, where he maintained bachelor quarters for some years. I often visited him during that time.

Floyd had a brilliant mind, a charming personality, and a fine legal education. He received many offers to go elsewhere to practice law where the monetary rewards would have been much greater, but he loved the New River area and its people with such intensity he could never divorce himself from them.

After the Appalachian Power Company, a subsidiary of the American Electric Power Company, a New York–based electric utility, sought a license from the Federal Power Commission to flood forty thousand acres of rich bottomlands in Ashe, Alleghany, and Grayson counties and drive three thousand people from their ancestral homes to enable it to supply peak-load electricity to faraway industries and people, Floyd enlisted my aid as a United States Senator from North Carolina and an old friend in his fight against the issuance of such a license. Unhappily, however, Floyd journeyed to that bourne from which no traveler returns before the fight was determined.

Thomas Schoenbaum details the events of the long fourteen-year fight by the people of Ashe, Alleghany, and Grayson counties to preserve for themselves and future generations one of the last free-flowing rivers in America—incidentally, a river which the geologists say is older than any other river on earth except the Nile.

In the fight to preserve New River, the people of Ashe, Alleghany, and Grayson counties were originally arrayed against an arrogant bureaucracy, the Federal Power Commission, whose only concern was the production of electrical energy, and an arrogant industry which was anxious to destroy the handiwork of God in order to supply peak-load electricity to its customers in distant places.

The absurdity of the industry's proposal is illustrated by the fact that the evidence before the Federal Power Commission revealed that the power company would use four units of electrical power to furnish only three units to its peak-load customers by dams which would fill up with silt and become useless after half a century.

Time after time the fight seemed to be hopeless as the Federal Power Commission and the courts made adverse decisions. After a time, however, the State of North Carolina, conservationist groups throughout the nation, and the nation's news media allied themselves with the people of Ashe, Alleghany, and Grayson counties. They were ultimately assisted by the arrogance of the Federal Power Commission and the arrogance of the American Electric Power Company, the real owner of Appalachian Power Company.

After the Senate had passed the bill to require the Department of the Interior to study the New River for possible inclusion in the National Wild and Scenic Rivers System and to postpone the construction of the proposed dams for a two-year period by the vote of forty-nine to nineteen, the Federal Power Commission without further ado granted a license to the Appalachian Power Company, but postponed its effective date from July to January. By so doing, the Federal Power Commission told the Congress, in effect, that they had only a few months in which to legislate. This arrogant act ultimately won many votes in the Congress for the preservation of the New River.

After the bill had passed the Senate, the Appalachian Power Company and its parent body, which were reputedly antiunion, made a deal with the national AFL-CIO by which they reputedly agreed to allow the union to control all jobs necessary to the construction of the dams for union help in defeating the New River bill.

Under the leadership of Roy Taylor, a representative from North Carolina, the House Interior Committee reported the bill by a substantial majority for House action, and it was apparent that the House itself was ready to pass the bill. Unfortunately, the House could not act on the bill unless the House Rules Committee permitted it to do so or unless it overruled an adverse action by the House Rules Committee by a two-thirds majority.

In consequence, the Appalachian Power Company and its parent body, with the aid of the national AFL-CIO and virtually all of the power utilities in the United States, urged the Rules Committee to refuse to allow the House to consider the bill. Since several of the members of the House Rules Committee were subservient to business interests and a majority of them had received contributions from labor during their last election campaigns, the House Rules Committee refused to permit the House to consider the bill, and while the House voted overwhelmingly to consider the bill, the proponents were unable to obtain the necessary two-thirds majority, and the year 1974 drew to an end with the prospect for saving New River in a desperate strait.

Subsequently, however, Governor James Holshouser of North Carolina was able to enlist the assistance of President Ford and the Department of the Interior in an effort to make the New River a part of the National Wild and Scenic Rivers System by administrative action. The prospect for success by this method was dimmed, however, by an adverse decision of the United States Court of Appeals for the District of Columbia Circuit. Hence, it became apparent that congressional action was needed.

Senator Jesse Helms of North Carolina and Representative Stephen Neal of North Carolina introduced companion bills in the Senate and the House to effect this objective and to nullify the license issued by the Federal Power Commission.

While these bills were pending, the American Electric Power

Company officials overreached themselves by testimony given before the House Rules Committee. By the testimony of its general counsel, Joseph Dowd, the American Electric Power Company stated, in substance, that the Appalachian Power Company would sue the United States for $500 million if Congress nullified the license issued to it by the Federal Power Commission. When the newspapers reported Mr. Dowd's testimony to this effect, I sent the following letter to each member of the United States Senate:

A newspaper clipping which undertakes to describe an event which occurred during the hearings held last week on this bill states, in substance, that Mr. Joseph Dowd, General Counsel of the American Power Service Corporation of New York, the parent company of Appalachian Power Company, testified that "they would sue the government for 500 million dollars" if the Congress of the United States dared to exercise its constitutional power and enact a bill prohibiting the construction of the dams on New River by Appalachian Power Company, which would flood thousands of acres of land in Alleghany and Ashe Counties, North Carolina, and drive several thousand people from their homes.

If Julius Caesar had heard this testimony prior to writing his book on the Gallic Wars, he would never have said that "All Gaul is divided into three parts." He would have said that "All gall is concentrated in this particular witness."

This is the first time in the history known to me when a commercial corporation undertook to intimidate the Congress of the United States, and frighten it into relinquishing its legislative powers under the Constitution. Surely, the Congress of the United States will not take this threat lying down.

I urge it to enact S. 158 without delay, and thus prevent these power companies from destroying the second oldest river on earth for the pecuniary benefit of corporations which attempt to intimidate the Congress of the United States.

Shortly thereafter the House passed the New River bill by a substantial majority. While the bill was awaiting action in the Senate, the supporters of the New River held a rally in the Senate Caucus Room in Washington. Hamilton Horton of Winston-Salem, a staunch advocate of preserving the New River, presided over this rally. Ham announced that the chief lobbyist for the Appalachian Power Company had entered the Caucus Room, and offered to extend to him equal time to address the rally. The chief lobbyist declined the invitation, and Ham Horton expressed the hope that the chief lobbyist would see a great light like Paul had seen on the road to Damascus and cease his lobbying efforts. I couldn't resist the temptation to inject myself at this point. I said, "Ham, I suggest that you inform those present that the great light which Paul saw on the road to Damascus came from the Lord and not from the Appalachian Power Company."

A short time later, the Senate passed the New River bill, and it became law when President Ford gave it his approval.

The New River was saved by the actions of persons beyond numbering. Ed Adams, Lorne Campbell, Wilmer Mizell, Steve Neal, Governor Holshouser, and Rufus Edmisten deserve a special mention. Thomas J. Schoenbaum, Norman Smith, and Millard Rich also merit much praise for the legal battles they fought in behalf of the New River, and Wallace Carroll did work beyond praise in alerting the news media and environmental organizations throughout the nation to the necessity of saving the New. Finally, Wilbur Hobby, president of the North Carolina AFL-CIO, merits commendation for breaking with the national AFL-CIO and supporting the preservation of the New River.

Sam J. Ervin, Jr.

Acknowledgments

MANY PEOPLE GAVE ME SIGNIFICANT HELP IN writing this book, and I should like to express my gratitude to them. Special thanks go to Dr. Ernest A. Carl, who coordinated the State of North Carolina's effort to save the New River, for providing me access to files and for recounting some of the incidents and stories that enrich this book. Dr. Arthur W. Cooper, now of North Carolina State University and formerly assistant secretary of the North Carolina Department of Natural and Economic Resources, provided many insights as well. I enjoyed working with the two other attorneys, Norman B. Smith of Greensboro and Millard Rich of the North Carolina Department of Justice, throughout the months the three of us represented the State of North Carolina in court.

Many people who live in the New River Valley provided help and encouragement. Edmund I. Adams of Sparta gave me access to his files and to those of Floyd Crouse and helped me understand the early phases of the fight against the Blue Ridge Project. Lorne Campbell of Independence, Virginia, and Sidney Gambill of Crumpler provided delightful company and hospitality and much useful material. Steve Douglas, who has a small farm in the valley, helped me understand the way of life he and others were fighting to protect.

Many people associated with the National Committee for the New River were a great help as well. John S. Curry, counsel to the Committee, and Joe C. Matthews, executive director, gave me access to their files and recounted much useful information. I am grateful to Elizabeth McCommon of Floyd, Virginia, for permission to reprint her beautiful "Ballad of the New River."

Finally, two grand old gentlemen, both legends in their own time, were kind enough to read over an earlier draft of the manuscript. Former United States Senator Sam Ervin of Morganton and

Professor Albert Coates of Chapel Hill gave me the necessary encouragement to see this project through to completion.

It must not be forgotten that it was the thousands of people who worked to preserve the New River and its way of life who really created this story. I only set it down to record their achievement.

Thomas J. Schoenbaum

Chapel Hill, North Carolina
August, 1979

Contents

Introduction

IT IS INCREASINGLY APPARENT THAT ONE OF the greatest problems confronting Americans during the rest of this century will be 'resource scarcity. The seemingly unlimited abundance of the continent that has sustained American civilization during the past two hundred years can no longer be taken for granted. Communities across the land are facing critical water shortages. Lumber and forest products are increasingly in short supply as easily accessible old-growth forests are diminished. Minerals necessary for the machines we rely upon or, in the case of uranium, for the generation of electrical power, are increasingly hard to find.

As a society we have entered a period of resource-scarcity inflation—prices for many of the goods and services we enjoy will rise relative to income as shortages of resources drive up costs of production. Many, if not all, Americans will have to settle for a lower standard of living and will undergo lifestyle changes as market forces reshape our national economy.

The shortage of energy is the most serious resource problem we face because energy is necessary for virtually everything we produce or consume. Our economy is now dependent on two non-renewable fossil-fuel energy sources, oil and natural gas. Domestic production of these fuels has peaked and will trend downward in the 1980s and 1990s. For a time the shortfall can be made up by imported oil and gas, but this too must decrease as the actions of international cartels, competition for scarce supplies, and our own balance-of-payments problem constrict our supply.

Thus the next twenty years will be a period of increasing reliance upon alternative energy sources. Coal and nuclear power have been expected to play a major role in supplementing dwindling oil and gas supplies. Yet there is serious question whether these energy sources, which now provide only 17 per cent of our primary energy production, can be expanded rapidly enough. Although the United States has an abundant supply of coal, long lead times are required to make it available, and there are unsolved problems of transportation, air pollution, and environmental degradation that will retard the growth of coal production. Nuclear power production also has not increased as rapidly as expected, as government and the public continue to grapple with the difficult issues of safety, nuclear proliferation, terrorism, and the disposal of nuclear wastes. Its future now seems uncertain.

There is clear indication, therefore, that the era of abundant, cheap energy is ending, and we are in a stage of transition to at least a temporary period of very high energy prices. This may result in a shift to a more labor-intensive and community-oriented economy and a growth in small-technology energy sources such as solar power systems. Our national tradition of substituting energy-intensive machines for labor will have to give way to some extent to the use of alternate technologies that emphasize human effort, energy efficiency, and the recycling of resources.

This change can be made without serious economic, political, or social disruption only if the transition is gradual and organic. Government policy should be directed toward facilitating the transition by encouraging not only the development of new sources of supply but also the more efficient use of existing energy resources. It is useless to try to prolong the era of relatively cheap energy by a supply-oriented, business-as-usual energy policy which attempts to meet existing demand regardless of the social, environmental, and long-term economic costs. In the first place, this policy will only buy a little time, ensuring that the period of scarcity, when it comes, will come suddenly, wreaking havoc on an unprepared economy. In the second place, the cost of unlimited energy development will be the disruption of traditional lifestyles and environmental degradation in many areas of the country. The short period of additional prosperity this will produce may not be worth the cost we will have to pay for it.

For two years, from August, 1974, to September, 1976, I was in the thick of a controversy between those holding this all-out approach to the development of energy resources and others holding the view that the costs of such development to the human environment were unacceptable. The stakes were the resources of the upper

New River Valley in the Blue Ridge Mountains of North Carolina and Virginia.

The protagonists of this controversy were, on one side, the American Electric Power Company, the nation's largest utility, and the Federal Power Commission, a federal agency whose mission it is to assure "abundant" supplies of electric power.* They planned to harness the electric power potential of the New River by building the Blue Ridge Project, a hydroelectric and pumped-storage electrical generating facility. (A pumped-storage facility generates electricity for consumption during periods of peak demand by using water that is previously pumped into a storage reservoir during off-peak periods.) This project was to have an installed generating capacity of 1,800 megawatts that would have produced 3.9 billion kilowatt hours of electricity annually; but to build it would have required the flooding of over 42,000 acres of the New River Valley.

On the other side were the majority of the residents and landowners of the New River Valley, who bitterly fought the project. Not only would many of them have been driven from their homes and businesses, but their whole way of life would have been radically altered. The most productive agricultural lands of the valley, the bottomlands of the New and its tributaries, upon which the farms depend for cash crops and winter feed for cattle, would suddenly have been gone. Established patterns of social intercourse would have been broken through the wholesale relocation of whole communities, roads, bridges, schools, and churches. In short, the living heart would have been torn out of the upper New River Valley, and the dislocations would have extended far beyond the actual project area.

My involvement with the controversy over the Blue Ridge Project began one day in August, 1974. Ernie Carl, then an environmental planner for the State of North Carolina, called me at my office at the University of North Carolina School of Law and said that the state had decided to appeal in the courts the order of the

---

* In 1978 the Federal Power Commission was renamed the Federal Energy Regulatory Commission.

Federal Power Commission licensing the project. He asked if I would help formulate legal arguments to present to the court. I readily agreed to work on the case.

At that point, I knew nothing about the project or the New River Valley except for a few articles in the *New York Times* written by Ned Kenworthy and a moving documentary on public television by Bill Moyers called "A Requiem for Mouth of Wilson," about a little town in Virginia that would be completely inundated by the project; but I was excited by the extraordinary prospect that a state government was about to take the lead in an environmental battle against a power company. Since I had often represented environmental organizations against the state, I looked forward to the novelty of joining hands with my former antagonists. I had no illusions that the state government had just experienced a conversion to the Sierra Club's philosophy; for North Carolina it was essentially a question of defending the state's rights and acting to prevent an out-of-state power company from appropriating a North Carolina resource for the benefit of customers located in other areas. Still, it was a rare spectacle and I was glad to have a part.

I could not foresee at that time the magnitude or the drama of the battle that was about to ensue. Before it was over, the New River Valley became a focal point of national attention; it was featured on national news programs and in editorials in hundreds of newspapers all over the country. The struggle was carried on in the courts, in Congress, and in the White House, and it became a major political issue in the North Carolina Republican Presidential Primary in March, 1976.

Even more remarkable is the fact that, in the end, the people of the New River Valley won their fight. In August, 1976, Congress passed a law prohibiting the construction of the Blue Ridge Project. The power company was forced to abandon its plans, and the people of the valley returned to their old way of life in a beautiful, out-of-the-way corner of Appalachia. The nation would have to forego the energy resources of the New River Valley.

This book is an attempt to explain how and why an energy-hungry nation made a conscious decision not to use an available energy resource and, in so doing, recognized the existence of American values and traditions that are even more important than energy development. Beyond that, the story of the New River is a study of how laws and political institutions operate and interact in our society. In the past decade we have, through legislation, recognized and given protection to the human, social, and natural environments. The laws that do so greatly complicate the ability of government and private industry to allocate resources for exploitative purposes. Would we be better off without such complications? The New River decision suggests that, although these environmental laws involve increased costs, to do without them would be more costly still.

Most of all, however, this book is about the thousands of people, residents of the New River Valley as well as other parts of North Carolina and the nation, who were able to take united action to overcome powerful corporate and labor union interests. It was a struggle of titanic proportions, but the will of the people prevailed. It is to them that this work is dedicated.

THE NEW RIVER
CONTROVERSY

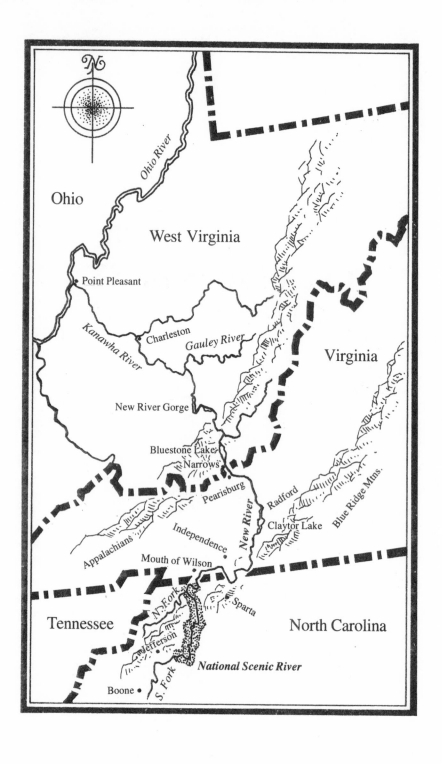

1

The River and the Valley: Geology and Prehistory

YOU MUST GO TO THE MOUNTAINS TO SEE THE New River. Travel west from the cities of central North Carolina, and the rolling, hilly countryside that is characteristic of the piedmont section of the state will end abruptly as you reach a high mountain barrier. This is the Blue Ridge, a long escarpment that runs through the state from the northeast to the southwest. Climb to the top and you will be standing on the Eastern Continental Divide, which separates the waters that flow into the Atlantic from those that seek out the Gulf of Mexico. In the northwest corner of the state, the stream draining the western slope of the Blue Ridge is the South Fork. A few miles west another, smaller stream, the North Fork, rushes down from the opposite ridge.

It has always seemed to me that these two watercourses flow in precisely the wrong direction. In this part of the country, rivers are supposed to run either southwest, between the parallel ridges of the Appalachians, or southeast toward the sea. Both the South and the North forks, however, head generally north toward Virginia, winding sharply in great bends and curves through a mountainous tableland with several peaks that exceed five thousand feet. After following their separate courses for over sixty miles, the two forks finally come together at a place appropriately called Twin Rivers, three miles south of the Virginia line. At this confluence begins the New River.

Continuing the northward journey, the New crosses into Virginia. Then, seeming to hesitate, it flows eastward and then south, again dipping down into North Carolina. After this show of reluctance to leave the land of its birth, the river gathers its strength and again strikes a course due north, reentering Virginia. At this point, incredibly, the direction of the New is almost at a right angle to the mountain ridges of the Appalachians. Paying no heed, it breaks the next ridge through a water gap in the Iron Mountains and enters the great central valley of the Appalachians, the Valley of Virginia.

Here the New winds among the population centers of Pulaski, Radford, Christiansburg, and Blacksburg. Near Radford, it is checked by an Army Corps of Engineers' dam to form an impoundment known as Claytor Lake. This, incidentally, is the part of the New River Valley that was the first to be settled by European immigrants in the eighteenth century.

The river does not tarry long among the cities of the central valley. Hurrying away from civilization, it again turns north, crashing through another mountain barrier, the high ridge of the Walker-Brush Mountains, the northern limit of the Valley of Virginia. Here the New finds itself in the beautiful, broad lowlands near Pearisburg, a great natural amphitheater surrounded by mountains. Soon it breaches another ridge, the Peters and East River mountains near a town called Narrows, to enter West Virginia.

The New River in West Virginia, after enduring another impoundment known as Bluestone Lake, courses northwestward in a deep, steep-sided valley known as the New River Gorge, one of the most spectacular natural features of the eastern United States. Here it is a wild river of dangerous white-water rapids. In central West Virginia, a few miles north of the New River Bridge, built to carry Interstate Highway 64 across the gorge, the New joins with the Gauley River to form the Kanawha. This river, the northward extension of the New, flows northwest past Charleston, West Virginia, finally emptying into the Ohio River at Point Pleasant. Thus, incredibly, the New-Kanawha system bisects the entire Appalachian range from the Blue Ridge to the Ohio River.

Whoever named the New River was apparently unfamiliar with

its origin and history.* Far from being "new," it is the oldest river in North America and one of the oldest rivers in the world. The New owes its longevity to its ability to maintain its channel through one hundred million years of mountain building and erosion.

The New River is geologically older than the Appalachian Mountains through which it flows. It originated during an even earlier period of mountain formation when, as scientists now believe, the African continental plate collided with eastern North America before the opening of the present Atlantic Ocean. This collision produced mountains higher than the present Alps of Europe. These were the geologic predecessors of the Appalachians, and they were drained by the ancestral New River. Known to geologists as the Teays, this river flowed north and west, emptying into the great inland sea that in prehistoric times covered the interior of the continent.

Over millions of years, the ancient Appalachians were eroded by rain, wind, and the action of the Teays and other rivers; the valleys between the mountain ridges filled with sediment until the region became a relatively flat plateau. At this time the Teays became what is known as a geologically *old* river. It no longer flowed swiftly in a relatively straight channel; because of the comparatively flat gradient it traversed, the Teays took on a meandering character with many bends and curves.

After several million years, new uplifting occurred, and the plateau was raised. Rivers formed which cut new valleys into the sedimentary rock formations of shale and limestone, leaving ridge crests composed of more resistant sandstone or conglomerate. Thus the stumps of the old mountains reappeared; these are the present Appalachians.

During this period, however, the Teays did not disappear but held

* Legend holds that the New River received its present name from the father of Thomas Jefferson, Peter Jefferson, who surveyed the valley in the eighteenth century. The most likely explanation for the name is that the New was the first river travelers encountered that flowed "behind the mountains," beyond the Eastern Continental Divide.

the course that had been established through the ancient plateau. Its waters took on new energy because of the uplifting, and they had the power to downcut the underlying formations, even the more resistant rock beds. In contrast to newer rivers largely controlled by the underlying surface, the Teays became, in geological terms, *superimposed* on the underlying rocks. It assumed the character of a *rejuvenated* river, entrenched within a valley while retaining its meandering course. This explains why its successor, the New River, exhibits the meandering characteristic of an old river, with multiple loops and bends, although it is a mountain stream in a deep valley.

During the time the Appalachians were being formed through this uplifting process, the Teays became what Professor Raymond E. Janssen has called "the master stream of a primeval America." Its headwaters were in what is now the central piedmont of North Carolina, which in prehistoric times was an eastward extension of the Blue Ridge. Millions of years of erosion by streams flowing east toward the Atlantic were later to wear down these mountains. But at that time the Teays flowed westward across them and entered the bed of the present New River. It traversed the present New and Kanawha river valleys to a point near Charleston, West Virginia. Here the Teays cut through a great valley to a point near Huntington, West Virginia, where it entered the bed of the present Ohio River. In southern Ohio it turned north to Chillicothe and Springfield, following for a distance the bed of the present Scioto River. Swinging west, the Teays intersected the present Indiana state line just south of Fort Wayne and continued west into Illinois, passing near the site of Champaign. Farther west, near Lincoln, Illinois, it was joined by the ancient Mississippi River, then one of its tributaries. Turning south, the Teays received the waters of another tributary, the ancestral Missouri, which drained the western plains. The mouth of the Teays was just south of the site of St. Louis, where it emptied into a narrow arm of the Gulf of Mexico, which then extended up the Mississippi Valley.

At this time, before the formation of the Ohio and Mississippi rivers as we know them, the Teays River was the principal river of the North American continent. The river and its floodplain were huge, estimated to be two miles wide between Charleston and Huntington and fifteen miles wide at the confluence with the Mississippi. During its long life it carried most of the sediment that now makes up the delta of the Mississippi.

The ancient Teays existed until the Ice Age, when great glaciers spread from the north into the North American continent. The ice covered the entire lower Teays Valley as far south as Chillicothe, Ohio, completely filling it with huge amounts of glacial drift. Below the edge of the glacier, in the middle valley of the Teays, the river waters were impounded into a series of finger lakes; the waters in these lakes gradually rose until they overflowed the lakes' rims. In this way a new system of rivers was created. The Ohio River was formed, usurping a portion of the Teays channel, and the Scioto and other southward-flowing tributaries north of the Ohio were formed to drain the higher elevations caused by the deposits of drift. The receding glaciers also produced the Great Lakes, and the Mississippi was established as we know it today.

The upper Teays River Valley was not covered by the glacier ice, nor were its waters impounded; it remained as a remnant of the once-mighty river. This is the present valley of the New River and its northern extension, the Kanawha River. No longer is the New the mightiest river of the continent; it is only a tributary of rivers that were once subservient to it. The unique history and the great antiquity of the New, however, explain why it is today the only river system to bisect the Appalachian plateau from east to west.

At some time during the period of great glaciations, known to geologists as the Pleistocene, man first appeared on the North American continent. The glaciers had locked up great quantities of water, lowering sea levels and producing a land bridge across the Bering Strait between Siberia and Alaska. This land bridge as well as a large area of central Alaska was never covered by the glacier ice, and it is known that many animals, including mastodons, mammoths, bears, bison, moose, and elk, migrated to North America from Asia during this age. Doubtless, bands of hunters from northeastern Siberia, modern man in the biological sense, followed the herds into Alaska. Estimates vary widely as to when this took place; they range from 10,000 B.C. all the way back to 40,000 B.C.

The first clearly defined cultural tradition in North America is called the Paleo-Indian or Big-Game Hunting tradition. It may have begun as early as 14,000 B.C. but seems to have flourished from 10,000 to 8,000 B.C. The Paleo-Indian was an undifferentiated culture that was centered in the western plains but predominated all across North America. Plant cultivation was not yet known, and the principal food supply was the large animals such as mastodons, camels, and long-horned bison, which then roamed North America. A nomadic way of life prevailed, and human society was organized into small hunting groups which followed the big game animals.

Tools evidencing this culture have been found in several sites in the New River Valley. The chief distinctive stone tool left by this culture is the Clovis Fluted projectile point, a leaf-shaped flaked point from $2\frac{1}{2}$ to $4\frac{1}{2}$ inches long with a short channel or flute at the base. Being larger than later projectile points, these are often found today in broken condition. It is believed these points were hafted to lances or spears.

The Paleo-Indian culture gradually diminished when, about the eighth millennium B.C., the prehistoric big game animals on which the Paleo-Indians depended became extinct. The giant bison, mammoth, mastodon, camel, and others mysteriously disappeared, either from climatic change as the glaciers finally retreated, or because, as some believe, of the depredations of early man.

A new way of life evolved as the populations of the big game animals declined. Men turned to more diverse food sources, principally smaller game animals, fish, shellfish, and wild plants. This hunting-gathering tradition produced more regional differentiation than did the Big-Game Hunting tradition, as populations adjusted to the primary food resources of their particular area. The Big-Game Hunting tradition was continued on the plains by the Plano culture, which was based on the bison. Elsewhere, however, new traditions began: in the west, the Desert tradition and in the east, the Archaic, which lasted until 1000 B.C.

The hunting-gathering lifestyle of the Archaic period required new types of tools and implements. Projectile points were smaller and were often stemmed, notched, or barbed. There were polished-stone woodworking tools such as axes, adzes, and gouges. Milling tools such as pestles and mortars were in use, as were stone vessels, ground slate points, and knives. Bone, horn, and ivory were used to make awls, needles, fishhooks, and harpoons. The people of the Archaic period were more sedentary, more numerous, and had more specialized skills than the nomadic Paleo-Indian hunters.

The New River Valley was first extensively occupied by man during the Archaic period, and the way of life of the inhabitants was probably very typical of the eastern woodland area at this time. Small occupation sites on the floodplain of the New River as well as on ridge crests imply that the valley was an important passageway through the Appalachians for early man. Certain sites may have been occupied seasonally, with the ridge crests used during the summer and the floodplain in the winter. Fish dams and weirs show that fish were a very important food source. Deer were also very important, not only for food, but as a source of sinew for cordage, hide for clothing, and bone and antler for a variety of tools.

During the approximately seven thousand years of the Archaic period, the early Indian cultures developed an increasingly complex hunting-gathering way of life. The implements and projectile points reflect this change, and archaeologists, through dating techniques, can estimate the millennium in which many distinctive styles of tools were made. Even though the New River sites have not been extensively excavated, the entire range of Archaic projectile points has been found. It has been established that many parts of the New River floodplain were continuously occupied throughout this period.

Around the first millenium B.C., another culture change began to appear in eastern North America and in the New River Valley. About 5000 B.C., the Indians of the central and southern highlands in Mexico had domesticated the wild grass now called corn or maize. They had also begun to cultivate a variety of other plants, such as squash, beans, and chili peppers. Gradually these foods became their dominant diet, and other plant resources were cleared away to allow large-scale cultivation of those plants on the river floodplains. Populations increased, and these Southern Indians established a largely sedentary way of life.

Over the course of many centuries, these developments in Mesoamerica were transmitted to eastern North America, apparently through the Mississippi Valley. Corn and other plants began to be cultivated on river bottomlands, and slash-and-burn agricultural techniques were developed. Clay pottery appeared along with ceramic figurines. The dead were buried in earthen mortuary mounds. This cultural tradition is known to archaeologists as the Eastern Woodland tradition; it began during what are called the Burial Mound periods. Pottery sherds that indicate the presence of this development have been found at numerous sites along the New River.

The Eastern Woodland tradition apparently began in the Ohio Valley with the Adena culture, which was succeeded by and incorporated into the Hopewell culture. Heavily influenced by the Mesoamerican developments, the Adena culture was established around 1000 B.C. by immigrants from the south. The Adenas had domesticated some plants but still relied primarily upon small game and wild plants as their food sources. The Hopewells, on the other hand, were skilled in maize cultivation. Pottery in both traditions was generally cord- or fabric-marked, i.e. with designs made in the soft clay, before firing, with a paddle wrapped in either a length of cord or a piece of cloth. Both peoples lived in small villages composed of a few small, round houses made of saplings.

The most evident remains of these peoples are the burial mounds they constructed for their dead. These are artificial earthen hillocks constructed over large, often log-lined pits in which the bodies were interred. The Hopewell culture developed a priestly cult of the dead, which is evidenced by artifacts found placed in the mounds.

The New and Kanawha River valleys constituted the southern fringe of this cultural area. Adena and Hopewell pottery has been found throughout these valleys, indicating the influence or domination of those cultures. It is believed that an extensive trade network existed along river valleys, and the New River Valley possibly served the Adenas and Hopewells as an important point of contact with the more primitive Woodland cultures of the southeast coastal region.

In contrast to the Adena and Hopewell traditions, most of the other peoples of the Eastern Woodland cultures subsisted on hunting, fishing, gathering, and slash-and-burn agriculture. They lived in small independent tribal groups and moved from place to place in search of food, often becoming involved in small wars with other tribes. The Indians the Europeans found living in the southern Appalachians and in the Ohio Valley, the Cherokees and the Shawnees, had become established there only in late prehistoric times.

The origin of the Cherokees is uncertain, but it is believed that this group, which spoke an Iroquoian language, came down from the north to settle in the area of the southern mountains, including the New River Valley. The Cherokees were perhaps a remnant group of mound builders of the Adena-Hopewell sequence of cultures who had been driven out by the Delaware tribes. This may have occurred in prehistoric times when the Delaware group occupied the Ohio Valley on their way to settling in Delaware, Pennsylvania, and eastern New York.

At the time of contact with the first Europeans, the Cherokees were living in the southern Appalachians in small, scattered, independent villages and did not have any well-developed tribal structure. The Spanish explorer Hernando de Soto passed through Cherokee villages in western North Carolina in 1540 on his expedition to the Mississippi River. Thus began the Cherokees' contact with Europeans. During the seventeenth century the English established trade relations with the Cherokees, and white traders lived among them. At this time an explicit hierarchy of tribal self-government was established and a Cherokee "emperor" was elected, apparently as a result of contact with European civilization.

The Shawnee tribes, who lived in the Ohio Valley and penetrated the New River Valley at the beginning of the historic period, were apparently a remnant group of the Delawares. They had come from the north after being displaced by the Hurons.

12/ THE NEW RIVER CONTROVERSY

The warlike Iroquois lived first on the shores of the St. Lawrence River and later settled in eastern New York State. They appear to have adopted an explicit policy of aggression against surrounding peoples, and in the late sixteenth century five of the Iroquois tribes agreed to form a league. In the succeeding decades these five "nations" sent out small armies to make war. They harassed the Shawnees of the Ohio Valley and sent war parties into Virginia and into the Tennessee and New River valleys, which probably checked Cherokee expansion to the north.

2

The New River
in the Old West:
Conflict and Settlement

BY THE TIME THE FIRST EUROPEANS VISITED THE
mountain lands west of the Blue Ridge in the seventeenth century,
the area had already seen intense human conflict. The Cherokee Indians, who lived in small villages along the Tennessee River and
its tributaries, had contested this area with the more northerly
groups of Indians, chiefly the Shawnees, who lived in the Ohio Valley and roamed throughout the middle-south region, and the Iroquois, who lived in present upstate New York. Bloody fighting
between the warriors of these rival groups had made the New River
Valley, as well as most of Kentucky, West Virginia, and southwestern Virginia, a no-man's land. Kentucky was referred to by the Indians as a "dark and bloody ground," and the middle Appalachians
were almost uninhabited, used only by roving bands as a battle-
ground and hunting area.

In 1642 Walter Austin, Rice Hoe, Joseph Johnson, and Walter
Chiles were granted permission by the Virginia House of Burgesses
to "undertake the discovery of a new river or unknown land bearing
west southerly" from the Appomattox River. This grant was apparently for the purpose of exploiting the possible mineral wealth
of the western lands, since one-fifth of the profits of any mines discovered was to go to the Crown. These men apparently never undertook the planned expedition into the mountains. They may have
been prevented by the Indian attacks that struck the Virginia colony
in 1644.

The threat of additional trouble prompted Governor William Berkeley to establish a series of back-country forts, including Fort Henry, at the confluence of the James and Appomattox rivers, near the site of present-day Petersburg, Virginia. Captain Abraham Wood, who had come over from England as a boy and had risen rapidly in the colonial militia, was given command of the fort. With Governor Berkeley's encouragement, Wood made Fort Henry an important colonial trading post and the departure point for expeditions to explore the country west of the Blue Ridge.

In 1650 Wood and his party set out through the wilderness of the Virginia piedmont and visited the area of the falls of the Roanoke River, at present-day Clarksville, Virginia, near the North Carolina line. This must have whetted Wood's appetite, for in 1653 he obtained a patent from the Virginia House of Burgesses and succeeded to the rights of Austin, Hoe, Johnson, and Chiles to enjoy profits from discoveries to the west "in places where no English ever have been." It is accepted by historians that in 1654 Wood mounted an expedition, crossed the Blue Ridge, and discovered the New River, although he left no written account of his expedition. He is believed to have traveled through what is now called Wood's Gap, in Floyd County, Virginia, first striking the river not far from the Blue Ridge near the present North Carolina and Virginia line. The New River was known as Wood's River in colonial times.

The first written account of the New River Valley was the result of the expedition of Captain Thomas Batts and Robert Fallam in 1671. They were sent by Governor Berkeley and Captain Wood in the hope of finding silver and gold, as well as a passage to the "great ocean" many were convinced was just west of the mountains. Batts and Fallam, accompanied by Indian guides, traveled up the valley of the Roanoke River, crossed the Blue Ridge near the present-day city of Roanoke, Virginia, and descended into the New River Valley. Here they noted initials carved into trees, a sign that they had been preceded by white traders or adventurers. They struck the New River, recognizing it as the westward-flowing river they were seeking as a passage to the sea. The party traveled down the river to about the present West Virginia line, where the New breaks through Peters Mountain, finding along the way tangles of old fields that had been under cultivation by roving Cherokee Indians. Before turning back, they set up a stick in the water and concluded that the water was ebbing and that the river was tidal! Their account reports that, upon ascending a hill and looking to the west, they saw "a fog arise and a glimmering light as from water." Thus their opinion that they were close to a great sea was seemingly confirmed.

Batts and Fallam took the occasion to claim the region on behalf of Charles II, king of England. They fired their guns and carved several trees with the insignia of the king and the initials of Berkeley and Wood. This claim became a legal basis for the English assertion of sovereignty over those lands of the North American heartland that are drained by the waters flowing into the Ohio River. Coincidentally, in this same year, the Frenchman Daumont de Saint-Lusson, asserted French sovereignty over all "countries, rivers, lakes, and streams . . . bounded on one side by the seas of the North and the West, and the other by the South Sea." Thus, as the seventeenth century drew to a close, the New River Valley was formally claimed by both England and France and was a rich hunting ground for the native peoples of the area.

White occupation and settlement were slow in coming to the New River Valley and to the other lands west of the Blue Ridge. For seventy years after the exploration of Batts and Fallam, this rich area was entered only by nomadic bands of Indian hunting parties and white fur traders. Early visitors saw an abundance of wildlife: ducks and geese were plentiful on the river waters and large mammals, including wolves, elk, deer, mountain lions, bears, and buffaloes, were common in the valley. Although virgin forests of chestnut, oak, hemlock, and hickory covered most of the area, much of the valley of the middle section of the New River was a shimmering grassland on which herds of buffalo roamed.

In the eighteenth century the Virginia colony tried to encourage the settlement of these new lands as a bulwark against the growing French power to the west. This policy was especially furthered by Governor Alexander Spotswood, an aristocrat who had fought un-

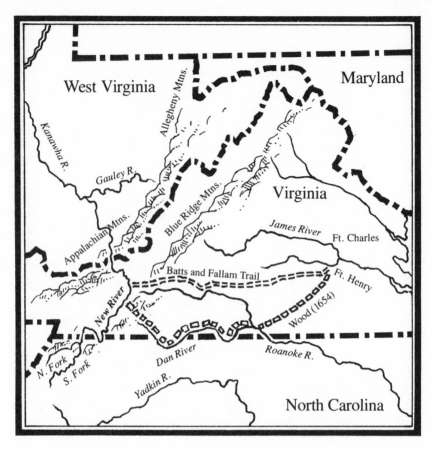

Probable route of Wood (1654) and Batts and Fallam trail (1671).

der Marlborough at Blenheim and who is chiefly remembered for his attempt to better relations with the Indians and for presiding over the construction of many of the beautiful public buildings in the Virginia capital, Williamsburg.

In 1716 Governor Spotswood decided to begin personally the opening of the west country. With sixty-three men and seventy-four horses, he traveled up the Rapidan River and climbed to the crest of the Blue Ridge. On this trip, true to his aristocratic position, he wore a plumed hat, green velvet riding clothes, and Russian leather boots. Once on top of the ridge, probably near Big Meadows in the present Shenandoah National Park, the gathering broke out the liquid refreshment—red and white wines, brandy, stout, rum, champagne, Irish usquebaugh, cherry punch, and cider—and drank to the health of King George and the royal family. Thus fortified, they descended into the Shenandoah Valley and camped overnight beside the Shenandoah River, which they named the Euphrates. Spotswood was convinced that this river flowed north into the Great Lakes. After a good dinner, they again toasted the health of the royal family, shouting the name of each member in turn and firing a volley after each name. This time they added the governor's health for good measure. Following his return home, Governor Spotswood named each of his companions to the order of the Knights of the Golden Horseshoe and gave each a small golden horseshoe inscribed *Sic Juvat Transcendere Montes* ("How pleasurable it is to cross the mountains!"). He touted the area as an agricultural paradise and referred to it as "World's End."

Governor Spotswood apparently persuaded few people by these antics. The opportunity to settle beyond the Blue Ridge was first taken not by Virginians but by German Palatine settlers coming down the Shenandoah Valley from Pennsylvania and New York. These families were members of the Lutheran or German Reformed churches and had left Germany to flee from religious persecution, wars, and exorbitant taxation. In 1726 Adam Müller built a cabin in the valley near the site of present-day Elkton, Virginia, and in 1731 Justus Hite settled near the present-day city of Winchester. By 1740 waves of German immigrants were coming down the valley, following the old Warriors' Path, the hunting trail used by the Iroquois and other northern Indians. German communities were founded in the New River Valley by Adam Harman (Heinrich Adam Hermann) at Horseshoe Bottoms, at the mouths of Thomas, Stroubles, and Back creeks, in southwestern Virginia. Settlement was encouraged by the low prices charged by the proprietor of these western lands, the fifth earl of Fairfax, who had inherited them from Lord Culpeper, the original grantee.

Scotch-Irish settlers also came down the valley in great numbers at this time. They were Presbyterians with a hostility toward both the English Crown and the Established English (Anglican) Church, looking for a new way of life on the frontier. The English and the Virginia colony were pleased to see them go to the back country, where they would be out of the way. One of them, Colonel James Patton, settled in the New River Valley at Draper's Meadow, on the site of the present-day Virginia Polytechnic Institute and State University.

German, Scotch-Irish, and other settlers also came down from the northern colonies into Virginia and North Carolina along the Great Wagon Road, which cut south through the back country of the piedmont east of the Blue Ridge. This was the route followed by Squire Boone, father of Daniel Boone, who came down from Pennsylvania to settle with his family in the Yadkin River Valley of North Carolina in 1753. One of these groups, the Moravian Brotherhood, German followers of John Hus and religious immigrants from a province in present Czechoslovakia, established Wachovia, an early permanent settlement in the back country of North Carolina. In the process they gave us the first recorded account of the upper New River Valley in North Carolina.

The story begins in 1752, when Lord Granville, the only Lord Proprietor of North Carolina who had not sold his lands back to the English Crown, began to look for settlers for his vast tract in the Carolina piedmont. He offered the Moravian Brotherhood, which in 1741 had established a settlement at Bethlehem, Pennsylvania, under the sponsorship of Count Zinzendorf, whatever land they wanted on which to found a community. Zinzendorf sent a group under the Moravian bishop Augustus Gottlieb Spangenberg to select an appropriate site.

Bishop Spangenberg and his companions first traveled south to Edenton, North Carolina, on the coastal plain, but found the inhabitants "lazy" and headed west. They traversed the piedmont, heading up the Yadkin River Valley, and coming to the Blue Ridge, decided to climb it, perhaps expecting to find an area similar to the Great Valley of Virginia. The ascent was made only with great difficulty on December 5, 1752, in the midst of a snowstorm. Over the ridge, Spangenberg's party came upon the twists and bends of the South Fork of the New River in what is now Ashe County, North Carolina. Instead of a broad valley, Spangenberg saw a mountainous tableland, drained by small creeks and the headwaters of the New. Although he pronounced the soil of the hollows and small valleys suitable for farming and grazing and explored the New River as far as Grassy Creek, Spangenberg rejected the area for settlement since he was looking for a large expanse of arable land. He returned east to the piedmont and decided to purchase instead 100,000 acres in what is now Forsyth County, North Carolina.

Spangenberg had made a fortunate choice in not settling in the New River Valley. A few years later the area became an arena of fighting in the French and Indian War, the struggle between the British and the French for domination of the North American continent. Early British defeats gave the Indian allies of the French confidence that they could push the English settlers out of the Old West. Although there had already been conflict with the Indians, especially around the settlements in the Holston and New River valleys, attacks began in earnest in 1755. On July 8 the settlement founded on the New River by the Inglis and Draper families was attacked and destroyed by Shawnee Indians. The inhabitants were either killed or captured, and Colonel James Patton, who was there on a visit, was also killed. The remaining settlers of the region took refuge in a rough stockade called Vause's Fort, where they were soon overcome by French and Indian forces. Other settlements in the Valley of Virginia experienced similar troubles. Many people were killed or withdrew to safety.

Governor Dinwiddie of Virginia and Governor William Dobbs of North Carolina responded to the French and Indian threat along the frontier by appointing commissioners to meet with the Catawba and Cherokee Indians, who had been relatively friendly, to enlist their aid against the French. The Virginia commissioners, Colonel William Byrd III and Colonel Peter Randolph, prevailed upon the Cherokee chief, Attakullakulla, whom the Virginians called Little Carpenter, to allow the building of a fort near Chota, on the Little Tennessee River, the chief town of the overhill (mountain) Cherokees. South Carolina also constructed Fort Loudon, about seven miles from the Virginia fort at Chota. Some joint military operations were conducted by the colonists and the Indians. In 1754 Major Andrew Lewis, who was in charge of building the Virginia fort, led a combined force of Virginians and Cherokees against the Shawnee Indians on the Ohio near the mouth of the Big Sandy River, near the site of present-day Huntington, West Virginia. This venture proved fruitless when they failed to find their enemies and provisions grew scarce, so they returned empty-handed.

Ultimately, however, the attempt at cooperation with the Cherokees proved to be a failure. Little Carpenter could not control the other Cherokee chiefs, especially the powerful Oconostota, who hated the English. Full-scale war broke out, and the British colonel Archibald Montgomery attacked and burned the Indian villages. The Cherokees responded by fleeing into the hills, waiting until Montgomery withdrew, and then overpowering Fort Loudon. But the Cherokee towns were again ravaged by Colonel James Grant, and the Cherokees were forced to make peace.

The English by this time had no further need for the Cherokee forts since they were prevailing elsewhere on the battlefield against the French. The Peace of Paris in 1763 ended the French North American empire and ratified the complete triumph of the British. Their sovereignty was confirmed east of the Mississippi River, and the New River Valley became British territory.

This treaty, however, marked the end of one conflict and the beginning of another for the western frontier. Although French claims were extinguished, festering grievances still remained between the

British and the Indian inhabitants. George III attempted to deal with these grievances by issuing the Proclamation of 1763, which reversed the pro-settlement policy. The claims of all the colonial land companies were nullified, and the western territories were declared to be a vast Indian reservation. Henceforth no lands could be purchased and settlement was not permitted west of the Eastern Continental Divide. The entire valley of the New River was thus reserved for the Indians. But even at the time of this proclamation, it was felt that the tide of settlement could not be stemmed so easily.

After the conclusion of the Peace of Paris, a few settlers remained on the New and Holston rivers, on the Indian side of the Proclamation Line of 1763. They had held firm against sporadic Indian attacks, and with the coming of peace, new families joined them. Influential men in the colonies who had obtained patents for large tracts for subdivision and resale to the settlers put pressure on the British to rescind or modify the proclamation. Dr. Thomas Walker of the Loyal Land Company ignored the anti-settlement policy and sold land along the Holston and Clinch rivers. George Washington and Benjamin Franklin, who had interests in land beyond the settlement line, predicted that the line would fall within a few years. An added problem was that the western lands were claimed and used as hunting grounds by three principal groups of Indians—the Cherokees, Shawnees, and Iroquois—each group the others' bitter enemy.

In response to these pressures, John Stuart, the British superintendent for Indian affairs in the Southern Department, negotiated the Treaty of Hard Labor with the Cherokees on October 14, 1768. This agreement set up a new settlement line running from Tryon Mountain, near the present border between North and South Carolina, northeast to Chiswell's mine, near present-day Wytheville, Virginia, and northwest to Point Pleasant at the confluence of the Kanawha and Ohio rivers. This treaty opened the entire middle and lower New River Valley to settlement. The upper New River Valley was to remain Indian territory.

Three weeks later, however, on November 5, 1768, Sir William Johnson signed the Treaty of Fort Stanwix with the Six Nations of

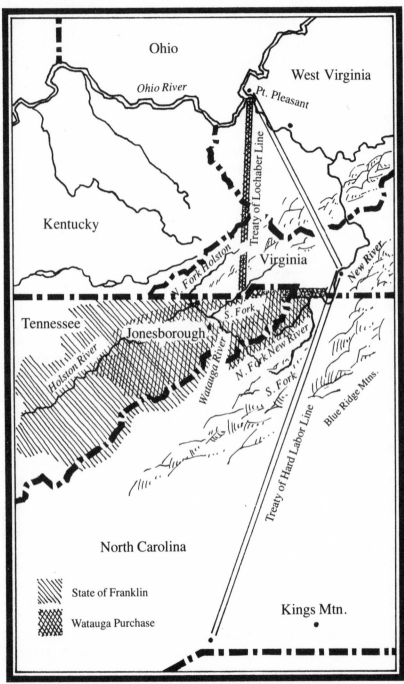

Map showing treaty lines, Watauga Purchase, and the State of Franklin.

the Iroquois confederacy, which, in return for allowing the Indians to keep their ancestral lands in New York and Pennsylvania, ceded to the colonies all the lands south of the Ohio as far west as the Tennessee River. One of the moving forces behind this treaty was Dr. Walker of the Loyal Land Company. News of this agreement infuriated the Cherokees, who did not recognize the Iroquois claims and regarded this treaty as inconsistent with the Treaty of Hard Labor. The Shawnees, who also claimed the Ohio Valley, were angered as well. The settlers on the Holston and the Loyal Land Company, on the other hand, chose to regard the Iroquois' cession as valid.

In order to settle this dispute, a new agreement was negotiated with the Cherokees at Lochaber, South Carolina, on October 18, 1770. A new line was drawn to take in all the white settlers in the Cherokee territory and to serve as a final line of settlement. The Lochaber line began at the point where the east-west North Carolina boundary (which had been run by William Byrd in 1729 and extended by Peter Jefferson in 1749) intersected the Hard Labor line of 1768; it ran westward to a point six miles east of Long Island in the Holston River and then directly north to the confluence of the Kanawha and Ohio rivers. This opened the way for a great influx of settlers in the triangle that now comprises most of southwest Virginia. But the upper New River Valley still remained outside the pale of settlement.

When the line was surveyed by Colonel John Donelson of Virginia in 1771, at Little Carpenter's urging, the Lochaber line was moved west to follow the Kentucky River to its mouth at the Ohio, thus including all of eastern Kentucky as well. This was accepted by Donelson on behalf of the Virginia colony, but the additional compensation promised to the Cherokees was never paid.

When the east-west line of the Treaty of Lochaber, which was the westward extension of the North Carolina boundary, was surveyed in 1771, however, it was found that a group of colonists who had established settlements and farms along the Watauga River (in present-day northeastern Tennessee) and the New River were in North Carolina, not in Virginia. This meant that they were in

Cherokee territory, since the westward settlement line in North Carolina was still the Hard Labor line of 1768. Virginia ordered the settlers to leave the lands, but the "Wataugans" refused. Since they were outside the protection of both the North Carolina and Virginia colonies, they negotiated directly with the Indians. In 1772 they leased their settlement lands from the Cherokees for about four thousand pounds in merchandise, household goods, and muskets. They also set up their own government, adopting a constitution known as the Watauga Association, which vested a court of five members with executive, legislative, and judicial powers.

The Wataugans' settlements were not the only incursion into Indian territory. Another group, known as the Long Hunters for the lengths of time they were absent from home, ranged throughout the wilderness west of the settlement line looking for adventure, game, furs, trade with the Indians, or land. Benjamin Cleveland of the upper Yadkin Valley in North Carolina led a small group that explored Kentucky from 1769 to 1772. Joseph Drake and Henry Scaggs organized a party from the New and Holston rivers and went through Cumberland Gap into the Kentucky country in 1769. Although these groups were trespassers on their lands, the Indians treated them rather leniently. Cleveland and his men were captured, made to give up their clothes and supplies, and sent home. A hunting camp left behind by Drake and Scaggs in the care of three men was attacked by an Indian named Captain Will, who took the 2,300 deer skins that had been accumulated and captured two of the men.

The best known of the Long Hunters was Daniel Boone, who had settled in the Yadkin Valley of North Carolina and ranged in his hunting expeditions over the mountains of eastern Tennessee and western North Carolina, including the upper New River Valley. Boone was restless. After Florida became British in 1763, he traveled into northern Florida to look it over as a possible place for his family. He found the climate wet and the game scarce, so he returned home. In 1769 Boone learned about Cumberland Gap and the Kentucky country from a friend, John Findlay, whom he had gotten to know when both served as wagoners in Braddock's ill-fated campaign of 1755. Findlay had traveled down the Ohio River and had visited a Shawnee town near present-day Winchester, Kentucky, where he learned of Cumberland Gap from some traders. Boone and Findlay, with some companions, immediately set out for Cumberland Gap and traveled up the Warriors' Path into Kentucky.

Boone eventually joined forces with the most prominent of the land speculators in North Carolina, Judge Richard Henderson. Henderson, who had financed Boone's exploring expedition of 1769, dreamed of making a fortune through the settlement of the Kentucky region. When his judgeship expired in 1773, he formed the Louisa Company, later called the Transylvania Company, and advertised choice lands for sale. In order to circumvent the treaty line of settlement, he opened negotiations directly with the Cherokees, inviting the aged leader, Little Carpenter, to Cross Creek (now Fayetteville), North Carolina. A private agreement for the sale of Cherokee lands was completed in 1775 at Sycamore Shoals, on the Watauga River. Henderson, encouraged by the Wataugans, purchased a vast tract from the Cherokees for about ten thousand pounds' worth of goods. This area included all of Kentucky west of the Kentucky River and most of the northern half of Tennessee, including the lands the Wataugans had leased from the Cherokees three years before. The latter lands were immediately reconveyed to the Wataugans for £2,000. These included Carter's Valley, and the valleys of the Holston, Watauga, and Nolichucky rivers in present-day Tennessee. Also purchased was the upper New River Valley in North Carolina.

Two Cherokee chiefs, Oconostota of Chota and Tsugunsin (known as Dragging Canoe), bitterly protested the sale. They predicted that it would threaten the survival of their people. Their judgment proved correct, for the 1775 Treaty of Sycamore Shoals started the chain of events that led to the forced removal of the Cherokees to the Oklahoma territory—the "Trail of Tears" of 1838.

These actions by the Transylvania Company and the pressures of the other land speculators were opposed by the British. The policy of the Board of Trade and Plantations was to encourage settlement of the newly acquired Canadian and Florida provinces as this would increase trade with the mother country. In February, 1774, the Crown ordered the colonial governments not to recognize any grants by the Indians and forbade sales of land west of the treaty lines.

Before news of this declaration reached the colonies, however, the Virginia land speculators, especially Patrick Henry, Thomas Wharton, and William Byrd, succeeded in gaining the support of the colonial governor, Lord Dunmore, probably by convincing him that he could profit personally from the western lands. One additional problem had to be solved before the lands of the lower New, Kanawha, and Ohio valleys could be opened for settlement. Although the Stanwix treaty had ousted the Iroquois, and the Lochaber line took care of the Cherokees, the Shawnees, who lived in the Ohio Valley, also claimed the area. The Long Hunters and land agents were killing off the Indians' game, and on several occasions engaged in unprovoked attacks against the Shawnees. When the Indians struck back, Dunmore ordered out the Virginia militia, thus beginning the single-battle conflict known as "Lord Dunmore's War." On October 6, 1774, a force led by Dunmore, Daniel Morgan, and Colonel Andrew Lewis defeated the Shawnees under Chief Cornstalk at Point Pleasant, where the Kanawha River joins the Ohio. In the ensuing peace negotiations, Dunmore, in defiance of the British policy, signed the Treaty of Camp Charlotte, under which the Shawnees ceded all their claims to the western territory. Although the Shawnees later contested the settlers in Kentucky, especially at the Transylvania settlement of Boonesborough, this battle effectively broke their power in the lower New and Kanawha valleys, opening the entire area to settlement.

On the eve of the Revolution, the entire New River Valley had been wrested from the Indians. The upper valley was the eastern fringe of the Watauga Association and the middle and lower valleys were part of Virginia. Settlers fanned out into the mountainous areas of the New in North Carolina and West Virginia after traveling down the Valley of Virginia from Pennsylvania and points north. These people, as well as the others of the old eastern frontier, were different from the earlier settlers in the Virginia and North Carolina colonies. They were Scotch-Irish and German farmers, who belonged to the Presbyterian and other, smaller, Protestant churches and generally did not hold slaves.

In Virginia the events leading up to the Declaration of Independence reached a climax on March 23, 1775, when Patrick Henry at the Virginia Convention in Richmond, which had been called to organize a colonial militia, made his "give me liberty or give me death" speech. This caused Governor Dunmore to seize the powder stored at Williamsburg and place it on a ship offshore. Henry immediately organized the militia and marched on Williamsburg to demand either payment or the return of the powder. Dunmore paid for the powder but two days later declared Henry an outlaw. News of the battles at Lexington and Concord in April, 1775, spread over the colony, and the Second Continental Congress met in Philadelphia in May.

These events had repercussions in the western settlements. Men from Fincastle, Virginia, gathered at the Lead Mines on the New River on July 15, 1775, expressed support for Henry, and condemned the British actions in Massachusetts. The county also organized a Committee of Safety and sent a contingent of riflemen to Williamsburg under Captain William Campbell.

In general, however, news of the outbreak of hostilities and creation of the Continental Army under Washington had little effect on the people of the frontier. These events were remote from their daily lives, they did not have a clear idea of what was happening, and they felt little or no loyalty to either the Virginia planters or the colonial government. Few frontiersmen answered the call to join Washington's army.

Passions were aroused, however, by news that British agents were urging the Cherokees to attack the settlers. The British assured the Indians that the old treaty lines were still valid and that the settlers would be forced to withdraw from the west. After several incidents, small stockades were erected by the settlers for protection against Indian attack. Old Fort Robinson at the Long Island of the Holston (near present-day Kingsport, Tennessee) was repaired and renamed

Fort Patrick Henry. Forts were also built at Watauga (present-day Elizabethton, Tennessee) and at Eaton's Station, about seven miles east of Long Island, on the path leading to Fort Chiswell on the New River.

In May, 1776, a letter purporting to be signed by Henry Stuart, deputy superintendent of Indian affairs for the British government, was delivered to Charles Robertson of Watauga. It stated that the Cherokees were now allies of the British and that a joint British-Cherokee expedition would be mounted to come up from the south and take possession of the entire west country of North Carolina and Virginia. The Wataugans were promised safety if they took an oath of allegiance to the Crown and joined the forces of the king. This did not accomplish its purpose, however, and the Wataugans appealed for help to the revolutionary forces in North Carolina and Virginia. They especially looked to Fort Chiswell and the Fincastle County militia for aid. Colonel Preston, the county lieutenant, sent lead shot from the mines of the New River Valley to the settlers, along with powder, and a few men; he also promised to take their case to Williamsburg.

The Cherokees under Oconostota and Dragging Canoe, ignoring Stuart's advice to wait, prepared to march north on the Watauga settlements with seven hundred braves. The settlers were given warning by Little Carpenter's niece, Nancy Ward, who was in love with a settler named Joseph Martin. The conflict was joined on July 19, 1776, at Island Flats, a level wooded area near the Long Island of the Holston. After a fierce, often hand-to-hand fight, the Indians fled when their leader, Dragging Canoe, received a serious wound. It is recounted that Robert Edmiston used profane language while repelling a furious assault by the Cherokees and that he was later brought to trial for his "crime" by the Ebbing Spring Presbyterian Congregation. Another Indian attack, on Fort Lee at Watauga, was also repulsed, but panic had seized many settlers, who fled up the main road and gathered at Black's Fort in Abingdon, Virginia.

Formidable help came from the revolutionary authorities of the colonies, however, which shortly broke the Cherokee power. In

July, 1776, Georgians under Major Samuel Jack burned several of the southern Cherokee towns along the Tugalo River. Troops from North and South Carolina under William Lenoir, Benjamin Cleveland, and Archibald Montgomery raided the middle Cherokee villages. Colonel William Christian of the Virginia militia, called home from the regular Continental Army, gathered 1,800 men at the Long Island of the Holston and burned Cherokee towns in the heart of the Cherokees' land, along the French Broad and Little Tennessee rivers. At Chota, the Cherokee chiefs surrendered, except for Dragging Canoe, who fled to the Tennessee River. His defiant band, which came to be known as the Chickamauga tribe, was not defeated until 1794.

In November, 1776, the Wataugans firmly joined their lot with the revolutionary forces. They renamed Watauga the Washington District, after George Washington, and four of their representatives, including John Sevier, participated in the Halifax Congress, which adopted the North Carolina Constitution on December 18, 1776. Washington County was organized to extend west from the crest of the Blue Ridge in North Carolina and include the upper New River Valley and all of Tennessee. Similarly, Fincastle County was reorganized on December 31, 1776, under the Constitution of the Commonwealth of Virginia into three large counties: Kentucky, Washington, and Montgomery. The central part of the valley of the New River was a part of Montgomery County.

The conquest of the Cherokees was ratified by two treaties signed by the Indians with the commissioners of North Carolina and Virginia on July 20, 1777, at the Long Island of the Holston. The grim-faced Oconostota, the aged appeaser Little Carpenter, Old Tassel of Toqua, and the Raven of Chota ceded to Virginia all their claims to present southwestern Virginia, while North Carolina got all of present northeastern Tennessee and northwestern North Carolina, including the upper New River Valley.

During the last years of the Revolutionary War, the South became a major theater of action. Georgia was taken by the British in 1779, and Charleston fell to Sir Henry Clinton and Lord Cornwallis on May 12, 1780, after which a Loyalist government was set up in

South Carolina. Cornwallis established a line of forts across northern South Carolina from Camden to the frontier settlement of Ninety-Six.

The American counterattack was led by General Horatio Gates, newly appointed commander of the Southern Department. He assembled an army of about three thousand at Hillsborough, North Carolina, and marched south to engage Cornwallis at Camden. After a fierce fight, Gates's army was thoroughly routed. Gates himself headed the "retreat," galloping on horseback sixty miles to Charlotte without stopping. It was later said that this was a most remarkable feat for a man of his age.

It looked then as if nothing could stop Cornwallis from taking North Carolina. He divided his army and dispatched Major Patrick Ferguson with an army of four thousand South Carolina Loyalists to march north on a route parallel to his own in order to secure his left flank. Ferguson's tactic was to frighten the western frontiersmen into joining his army. He laid waste much of the countryside, burning homes and stealing cattle. He called the mountain men barbarians, mongrels, and backwater men, appealing to the Tories of the frontier to join him, lest they be "pinioned, robbed . . . murdered" and "pissed on forever" by the mountain men.

This brought an angry response. Local Tories who attempted to join Ferguson were prevented from doing so whenever possible. In the upper New River Valley of North Carolina, Americans under Captain Robert Love defeated one such group of 150 Tories in July, 1780, in the Battle of Big Glades, near Old Fields Creek. The pioneers of the Nolichucky, Holston, Watauga, Clinch, and New River valleys assembled at Sycamore Shoals under John Sevier, William Campbell, and Isaac Shelby. They crossed the Blue Ridge at Gillespie's Gap and picked up Benjamin Cleveland and his men at Quaker Meadows. Ferguson made his stand on October 7, 1780, on top of Kings Mountain, which straddles the North and South Carolina line. The frontiersmen overwhelmed the Loyalist force, killing Ferguson and all who did not surrender; it was a key victory for the American side.

In the unsettled period between the end of the Revolutionary

War and the ratification of the Constitution of the United States in 1789, one more political movement occurred that affected the people of the New River Valley. This was the formation and the dissolution of the State of Franklin.

The Continental Congress as early as 1780 had adopted the basic principle that western lands ceded to the Confederation would eventually be eligible for statehood as equal and independent states. In 1784, after Virginia ceded her lands north of the Ohio, North Carolina ceded her lands west of the Blue Ridge on the condition that they be accepted by Congress within twelve months as counting toward her share of the expenses incurred during the Revolution. The state legislature also wanted to be rid of the frontiersmen of the New and Watauga rivers, whom many of the legislators regarded as the "offscourings of the earth." The people of these western lands responded by holding a convention at Jonesborough, in what is now Tennessee, to make plans for eventual statehood.

Before statehood could be declared, however, the North Carolina legislature in October, 1784, took back these western lands by repealing the act of cession. It organized the area as the new judicial district of Washington, placing it under military rule, with John Sevier as Brigadier-General. In 1785 the people of the region resisted North Carolina's action, however, and adopted a constitution at Jonesborough setting up a new government, called the State of Franklin. They selected delegates to be sent to the Continental Congress and elected Sevier as the first governor.

The State of Franklin collapsed within four years, however. The Continental Congress refused to recognize it, Virginia was hostile because it feared an attempt to annex its southwestern counties, and North Carolina regarded it as a "mock government." In 1789 North Carolina again ceded the western lands to the national government, this time retaining that part of the State of Franklin which included the upper New River Valley and present Ashe and Alleghany counties. The western lands ceded were admitted to the Union in 1799 as the new State of Tennessee. Thus the upper New River Valley very narrowly missed belonging first to the State of Franklin and then to Tennessee, instead of North Carolina.

3

The Lost Provinces

WITH THE PASSING OF THE FRONTIER WARS, THE
New River Valley entered a period of relative isolation from the
rest of the nation. Because of poor roads and poor communications,
the inhabitants were left alone to develop an agrarian, mountain
way of life. The small, family farm became the basic economic unit
of the area.

The central portion of the valley, where the New River meanders
across the great central Valley of Virginia between the Blue Ridge
and the Allegheny Mountains, remained an important link in the
settlement of the lands beyond the Appalachians. The famous Wilderness Road led through the New River settlements, and a ferry
was operated to carry travelers across the river at a place known as
Ingles Ferry. Thousands of pioneers funneled across the New River
on their way toward Cumberland Gap and Tennessee and Kentucky beyond. Settlements such as Abbeville and Han's Meadow,
later known as Wytheville and Christiansburg, grew up as stopping
points along the Wilderness Road. But the population of the area
grew slowly; only with the building of a railroad in 1856 were towns
such as Radford and Blacksburg incorporated.

Railroads eventually put an end to the isolation of the lower
valley as well. The New River Gorge region of present West Virginia remained a wilderness until 1873, when the Chesapeake and
Ohio Railroad Company completed its line up the New and Kanawha river valleys to the Ohio at Huntington. This opened the
valuable coal fields of the area, and small towns sprang up along the
railroads and creeks to accommodate the needs of the railroad and
mine workers. Life in these camps was boisterous and conditions
were primitive.

A different way of life characterized the upper New River Valley, however. The present tri-county area of Grayson County, Virginia, and Ashe and Alleghany counties in North Carolina comprises a unique, unified mountain region, very different from the steep-sided gorges of the lower valley or the broader middle Valley of Virginia. The winding courses of the New River and its tributaries enclose rich bottoms of seventy-five to two hundred acres against the curve of the hills, creating an almost perfect setting for small farm units. This geographic characteristic was the determining factor in the cultural and economic evolution of the area.

In 1793 Virginia established Grayson County (including present Carroll County), named after Colonel William Grayson, one of the first two United States senators from Virginia; in 1799 North Carolina established Ashe County (including present Alleghany and Watauga counties), named after the state's governor at that time. Both states began to make grants of land for homesteading to veterans and others in order to open the mountains to settlement. Families filtered into the region to establish farms in choice areas. Several grants were also made to foster exploitation of the iron ore deposits found in the hills above the river. The Ashe County court records a grant in 1807 to one Daniel Dougherty to set up a forge and iron works on Big Helton Creek. Another grant of 3,000 acres was made to Thomas Calloway in the same year for the production of 5,000 pounds of iron. Copper mining also became important during the nineteenth century; Ore Knob Mine produced iron and copper for over a hundred years.

Agriculture was, without question, the major economic activity in the valley, however. The earliest buildings constructed along the river were log cabins to provide immediate shelter for the settler, his family, and his animals. The most common type was the double pen or dogtrot, which consisted of two rectangular pens separated by a passage. The family lived on one side, while the animals and foodstuffs were housed on the other.

Typically, as farm productivity and family size increased, a settler would construct a new log house and various outbuildings, creating a farm complex. The older log building was then used as a barn or

A family log house in the Helton vicinity of Ashe County. (*Courtesy of Department of Archives and History, Raleigh, N.C.*)

The Greer-Parsons house in Grassy Creek, Ashe County, constructed between 1812 and 1817. (*Courtesy of Department of Archives and History, Raleigh, N.C.*)

The Walter Greer house, typical frame house with ornamental woodwork, Ashe County. (*Courtesy of Department of Archives and History, Raleigh, N.C.*)

The Robert Livesay house, Grassy Creek vicinity, Ashe County. (*Courtesy of Department of Archives and History, Raleigh, N.C.*)

Two views of the John Jones house in Grassy Creek showing the rolling hills typical of the valley. (*Courtesy of Department of Archives and History, Raleigh, N.C.*)

granary. The second-generation log houses, many of which still survive, were usually of half-dovetail construction, built of hand-hewn logs with exterior chimneys of dry-laid or mortared stone. Surviving structures of this type rest on full stone foundations; they range in form from one story, to a story and a half, to two full stories.

By the middle of the nineteenth century, existing log houses were often improved with frame additions and weatherboarding. A traditional improvement was the addition of a one- or two-story frame "L" at the back of the house. Windows were often replaced, and new ones were cut through the walls. Board-and-batten doors were replaced by three- or four-panel doors, and partitions were often constructed to create smaller rooms.

A few prosperous families built brick houses in the upper valley. The oldest example of this is the Greer-Parsons House in Grassy Creek, a farming community in northern Ashe. The house, a three-bay, two-story building, was constructed by slave labor between 1812 and 1817.

The owner, Aquilla Greer, had the largest farm in the area. His sons and daughters remained on the land, and individual farms multiplied to form a prosperous agricultural community that shared common boundaries, machinery, and other resources. They raised Shorthorn cattle for meat and milk, and crops which included corn, wheat, hay, and vegetables. The 1850 census of Ashe County lists the property of John Greer, Aquilla's son, as including 17 slaves, 225 cleared acres, 465 acres of woodland, 9 horses, 17 milch cows, 4 oxen, 34 head of cattle, 53 head of sheep, and 46 hogs.

Grassy Creek was the most prosperous of several concentrations of settlers throughout the upper valley and along tributary creeks. Other communities were formed by the Hamilton family at Beaver Creek and the Duvall and Perkins families on Helton Creek. Although many families held slaves, theirs was not a planter society based on the exploitation of black labor. The 1860 census gives the population of Ashe County as 7,956, including 391 slaves and 142 free Negroes.

The outbreak of the Civil War divided the people of the New River Valley. Many opposed secession, and eighteen sons of Ashe

County enlisted in the Union Army. The split was felt in the churches; many congregations of Baptists and Methodists divided along political lines. The majority of the people went along with their states, however, and several hundred recruits joined the Confederate Army.

Though the valley supplied the Confederates with iron, lead, and agricultural products throughout the war, its isolation protected it from the war's excesses. The biggest problem seems to have been the large number of deserters and others who took to the mountains and raided the farm families for food. The only actual fighting in the area consisted of isolated raids by Federal troops. One of these was directed against the lead mines on the New River in Virginia, and in 1865 Colonel George H. Stoneman sacked the depot of the Virginia and Tennessee Railroad at Christiansburg.

After the war, the agricultural economy of the valley continued to prosper. Most of the families were third and fourth generations of the original pioneer families of the valley: the Gambills, the Greers, the Blevinses, the Neavses, the Reeveses, the Sturgells, the Phippses, and the Waddells. There was a gradual transition from subsistence farming to cash crops, and corn and wheat became increasingly important. During this period, large, Victorian-style frame houses were constructed, many of which still ornament the valley.

The Grassy Creek Community is the best example of these large farm complexes constructed between 1885 and 1920. The area is dotted with frame houses sheathed in weatherboard with two stories in either a "T" or an "L" plan. Ornamental woodwork sets off the porches, eaves, gables, windows, and doors. Outbuildings include springhouses, root cellars, meat houses, woodsheds, and large barns covered by gable or gambrel roofs.

Grassy Creek prospered until the eve of the First World War. In 1915, following the unwise advice of the North Carolina Department of Agriculture, the Greer family decided to develop a commercial cheese-making industry in the valley. Their government advisers in Raleigh considered the area ideally suited for dairying operations and the export of cheese to population centers in the

North. This attempt to depart from the traditional farming of the valley proved disastrous. Roads were still primitive, and transporting the cheese out of the valley was too difficult. By 1920 the experiment had failed, and most of the Greer family sold their farms and moved away, leaving the valley to less ambitious small-farm families.

The commercial and industrial ventures that succeeded in the valley were those geared to the local economy. Carding mills were operated on many streams, including Dog Creek and the North Fork near Warrensville. Gristmills and woodworking plants took advantage of the abundant water power of the river and turned out products for the local farmers. Field's Manufacturing Company, a woolen mill founded in 1884 near Mouth of Wilson, Virginia, is still in operation.

In July, 1883, mineral springs were discovered near Crumpler, in Ashe County. According to local tradition, Willie Barker, a son of Eli Barker, was helping his father plow a hillside one day and went to get some water to drink. His father, after tasting the water, pronounced it the best he had ever had. Willie's hand and arm were covered with sores from poison oak, but the next morning they were nearly healed. The next day his father, also suffering from poison oak, bathed in the water and was cured. The word spread, and soon hundreds of people were coming to take advantage of the find.

In 1885 a Captain Thompson of Saltville, Virginia, purchased the springs and began to construct a summer resort at the site. He called it Thompson's Bromine and Arsenic Springs Hotel and began an advertising campaign to attract customers. The waters were called a "medicinal beverage performing miraculous cures" and "the most remarkable discovery of the nineteenth century."

The campaign was an immediate success, and one of the results of the demand for the spring water was the building of the first road from the upper valley to the outside world. In 1887 convict labor was used to construct a wagon track from Jefferson to Marion, Virginia, by way of Thompson's hotel. Once the road was constructed, an average of fifty wagons per day hauled water from the springs to the railroad at Marion. Thompson's Bromine and Arsenic Hotel

Thompson's Bromine and Arsenic Spring, Crumpler vicinity, Ashe County. *(Courtesy of Department of Archives and History, Raleigh, N.C.)*

was continually in use until it burned down in 1962, but the spring-house still stands with a sign proclaiming the "genuine bromine and arsenic springs" inside.

Other mineral springs were discovered in the area. Among these, Shatley Springs, about four miles from the site of Thompson's, became the best known. Its discoverer, Martin Shatley, soon spread the word of the springs' healing powers and offered the water free of charge to all who wanted it. He also built a small rustic resort to provide the visitors with cabins and fresh mountain food. This resort became very popular and continues to operate today.

In spite of the popularity the mineral springs acquired, at the beginning of the twentieth century the New River country was still an almost unknown appendage to the rest of North Carolina. Because of its isolation and its dependence upon the railroad in Marion, Virginia, the area was known as the "Lost Provinces." There was no connection by road or rail to the piedmont cities of North Carolina.

In 1900, however, Charles Brantley Aycock visited Ashe County while campaigning for governor. In a speech in Jefferson, he declared that "Ashe County should have a road letting it out at its front door." After his election he fulfilled his promise by providing convicts from the state prison to grade a dirt path up the Blue Ridge, which was grandly named the North Wilkesboro–Jefferson Turnpike. This road was a false hope, however, since erosion and mud made it impassable most of the time. It was finally closed completely by the floods of 1916.

The next link with the outside world was built as a result of a decision by the Norfolk and Western Railway to exploit the forests of the New River region. In 1914 a new line was built from Abingdon, Virginia, into Ashe through Elk Cross Roads (now Todd). The railroad company was unwilling to put the line through to Jefferson, the county seat, with a population of about five hundred; to save money, the terminal depot was built a few miles away. The town of West Jefferson grew up around the railroad depot. In the next few years the great forests of oak, poplar, maple, walnut, hickory, and pine were unmercifully stripped and taken out through

Abingdon. The sawmills of the area worked constantly, and within a few years denuded hillsides signaled the end of the boom. The tracks between West Jefferson and Todd were summarily torn up, as the railroad was not interested in providing passenger service after the depletion of the forests. Once again, the area stood in isolation. The resiliency of the mountains was such, however, that second-growth forests soon began, and today the land has healed much of the scarring caused by the rape of the forests.

The long-awaited transportation link to the rest of North Carolina came only in the 1920s. Governor Cameron Morrison decided it was time to give Ashe County an all-weather highway, and State Road 16 was constructed up the Blue Ridge to Jefferson and West Jefferson. Today this is still the road to the valley from piedmont North Carolina.

During the decades of the 1940s and 1950s, other changes came to the New River Valley which benefited the people without seriously affecting their traditional way of life. The rural electrification program of the Roosevelt administration brought electricity into the farms and homes of the area, and small manufacturing plants were built, which contributed a balance to the economy of the family farm. The Oak Flooring Company was established in 1935, and the Peerless Hosiery Company began operating in West Jefferson in 1953. Other industries included the Phoenix Chair Manufacturing Company, the Kraft-Phoenix Cheese Manufacturing Company, the Sprague Electric Company, and the P. H. Hanes Knitting Company.

By the 1970s a stable prosperity prevailed in the valley. Although per capita income remains relatively low, averaging just over $2,000, this is misleading because people grow or build many of the things they need. Economic activity still centers around family farms, many of which have been in the same family for generations. Major cash crops in the region include tobacco, beans, and Christmas trees. Since the non-arable uplands are ideal for grazing, beef cattle and dairy farming are important. In Alleghany County alone, fourteen new dairies were established in the 1970s, bringing the total number in the county to seventy-six. Ashe County leads the state in beef cattle production. The lowlands along the river, as well as the uplands, are essential to raising cattle. The corn grown in these bottomlands, together with two cuttings of hay per year, provides winter fodder for cattle when the winter snows prevent grazing on the uplands.

Necessary community services, such as churches and schools, have been provided through community effort and labor-in-kind. For example, the elementary school of the Ashe farming community of Healing Springs was built by farm families around 1920. The county provided the bricks, which at that time cost $1,000, and each family of the community was responsible for the construction of one school room. This building, still carefully maintained, is in use today.

There are no urban areas in the upper New River Valley; the principal towns, Jefferson and West Jefferson in Ashe, Sparta in Alleghany, and Independence in Grayson, ranging in population from one to three thousand, are service centers for the rural population. There are also numerous community centers, each consisting of several houses, a gas station, a general store, a church, and the like. But the region is essentially a high plateau of woodlands and pasturelands, interspersed with mountains and the river. One can still climb Mount Jefferson in Ashe County, as the scientist Elisha Mitchell did in 1827, and look out upon, in his words, "an ocean of mountains"; now there are cultivated valleys and small towns out there as well, whose soft beauty produces an overwhelming feeling of harmony between man and nature.

Testimony to the vitality of the traditional lifestyle here is the lack of any massive out-migration of people to urban centers. Although some loss of population occurred in the 1930s and 1940s, the decline had been stemmed by the 1960s and 1970s because of new economic opportunities. Small manufacturing plants provided several thousand factory jobs for the people of the valley without adversely affecting the traditional agricultural economy. The population of Alleghany County grew from 7,734 to 8,134 in the decade 1960–1970; Ashe County's population declined slightly from 19,768 to 19,571. People in search of higher-paying technical employment and a different lifestyle still must go to urban areas, but the area provides ample opportunity for the younger generation to enter

Canoeing the New. (*Courtesy of the* Winston-Salem Journal & Sentinel)

the deep-rooted culture of their parents. In the summer of 1976, the unemployment rate in Ashe County was 5.8 per cent and in Alleghany County, 2.2 per cent, compared with the national average of 7.9 per cent.

Though the valley retains its traditions, it receives a portion of its income from tourism and outside development. People still come to Shatley Springs to drink or bathe in its waters, which are still credited with miraculous cures and dispensed free of charge. Several other family-owned resorts also operate in the area. Real estate purchases and development by outsiders have added to the area's economy. Developments have been built near the Blue Ridge Parkway, and houses have been built on marginal farmlands as vacation homes. And, of course, those seeking recreation come yearly in large numbers to the river itself for its excellent white-water canoeing and fishing.

A trip down the New River in a canoe is a magnificent experience. Stretches of flatwater alternate with white-water rapids as the river winds past low hills with cultivated bottomlands or steep cliffs. Kingfishers and herons are common sights, and in the spring there is a chorus of birdsongs. In the early summer the canoeist sees the colors of the flame azalea, rhododendron, and mountain laurel, which appear on the heels of the earlier-blooming trillium and wild iris. In the fall the scene is dominated by the rich red, gold, and brown of the hardwood forests and the masses of yellow goldenrods on the riverbanks.

Despite the human prosperity of the valley, the New River ecosystem has remained relatively unaltered, and its waters are still unpolluted. Sixty-eight species of fish have been identified, including several species of darter and minnow that are rare or endangered.* The smallmouth bass is an important game fish of the area,

* According to the Department of the Interior the following rare or endangered species are found in the New River Valley: the aquatic snail, *Spirodon dilatata*; the New River shiner, *Notropis scabriceps*; the Kanawha minnow, *Phenacobius teretulus*; the bigmouth chub, *Nocomis platyrhynchus*; the tongue-tied minnow, *Parexoglossum laurae*; the bluntnose minnow, *Pimephales notatus*; the Kanawha darter, *Etheostoma kanawhae*; the blackside darter, *Percina maculata*; the sharpnose darter, *Percina oxyrhycha*; and the flathead catfish, *Pylodictis olivaris*. Despite the presence of these endangered species, however, their protection was never a major issue in the fight

and the tributaries of the river provide excellent habitat for trout.

About half the land area of the valley is covered with second-growth forests, and three distinct forest types are represented. The uncultivated bottomlands are sites for remnants of the Southern Bottomland Hardwood forest, including beech, maple, sycamore, willow, hickory, ash, and oak. On the sloping uplands, the Cove Hardwood forest is found. This is magnificent even as a second-growth forest, and beech, black gum, yellow poplar, hemlock, yellow birch, maple, oak, hickory, and ash are all common. The understory of this forest consists of dense thickets of beautiful flowering shrubs, such as mountain laurel, wild azalea, rhododendron, and viburnum. The exposed mountains and ridge tops are the domain of the third type of forest, the Oak-Hickory Hardwood type, which includes, in addition to oak and hickory, white and yellow pine, maple, and eastern red cedar.

---

against Blue Ridge. The New River controversy focused throughout on human concerns.

4

"Progress" Comes to the New River: Phase One of the Blue Ridge Project

IN 1962 APPALACHIAN POWER COMPANY, A SUBSID-
iary of American Electric Power Company, the nation's largest
electric utility, decided to look into the feasibility of generating
hydroelectric power on the upper New River. To do this, it had to
get the approval of the Federal Power Commission (FPC), which
must by law give permission before any nonfederal hydroelectric
project can be built. On March 11, 1963, the FPC granted Appalachian a permit to carry out a two-year feasibility study.

As a result of this study, on February 27, 1965, Appalachian
filed an application with the FPC for permission to build a two-
dam hydroelectric and pumped-storage facility, which it named the
Blue Ridge Project. A pumped-storage facility differs from a con-
ventional hydroelectric dam in that double storage reservoirs must
be built. Power is generated at times of peak demand by releasing
water from an upper reservoir to a powerhouse at the river; then,
during times of slack demand, water is pumped from the lower reservoir to the upper one, utilizing power from elsewhere within the
system. The proposed project was to have a total installed capacity
of 980 megawatts. The upper reservoir was to have a surface area of
16,600 acres, with a lower reservoir of 2,850 acres.

This plan did not arouse much opposition in the valley. Appa-

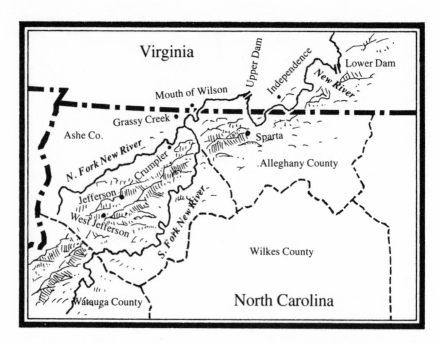

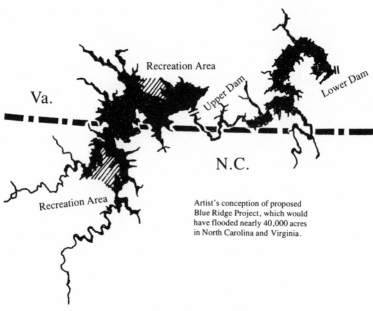

Artist's conception of proposed Blue Ridge Project, which would have flooded nearly 40,000 acres in North Carolina and Virginia.

lachian worked hard to convince community leaders that it would bring substantial economic and recreational benefits to the area. Most people felt that electric power represented progress, for which they should be grateful. Besides, it seemed useless to resist. A few people who were directly affected, such as Jack Phipps, who runs a dairy farm near Mouth of Wilson, Virginia, and Glenn Halsey and Jim Todd of Independence, wrote a series of protest letters to their political representatives but got no response.

Before the FPC could finish considering Appalachian's plan, in June, 1966, the U.S. Department of the Interior asked to be heard. Interior, which at that time was responsible for cleaning up pollution in the nation's waterways, was concerned about the wastes being dumped into the Kanawha River by chemical plants and other industries near Charleston, West Virginia. It wanted Appalachian's twin reservoirs to include enough storage capacity to be able to release water downstream during times of low flow so that the New and Kanawha rivers would be able to dilute the pollution.

When hearings were held on the project by the FPC in 1967, it soon became evident that a much larger facility would be needed if Interior's demand for water-quality storage was to be met. Accordingly, in June, 1968, Appalachian submitted a revised proposal that would approximately double the size of the original facility. The Modified Blue Ridge Project, as it was called, would have an upper reservoir of about 26,000 acres and a lower reservoir of about 12,390 acres. Each development was to include a powerhouse, transmission lines, a spillway, and other facilities. The upper dam was to be 300 feet high and 1,500 feet long; the lower was to be 236 feet high and 2,000 feet long.

The two developments together required the use of 27,900 acres of land in Grayson County, Virginia (9.6 per cent of the county area); 5,800 acres in Alleghany County, North Carolina (about 4 per cent of the county area); and 8,400 acres in Ashe County, North Carolina (about 3 per cent of the county area). Displaced from the project area would be 893 dwellings, 41 summer cabins, 10 industrial establishments, 23 commercial facilities, 5 post offices, 15 churches, 12 cemeteries, and 8 miscellaneous structures. It was estimated that at least twenty-seven hundred inhabitants would be forced to leave their homes. By the calculation of the Agricultural Extension Service in the affected counties, the loss in crop and livestock sales would total $13,500,000 annually (at 1973 prices), although this figure was sharply disputed by the power company. In any case, some forty-four miles of the main stem of the New River, twenty-seven miles of the South Fork, and twenty-three miles of the North Fork would be inundated, flooding hundreds of family farms and the richest agricultural lands in the upper valley.

Balanced against this, the project would initially generate about 3,900,000 megawatt-hours of electricity annually. The cost of the production of this amount of electricity with coal-fired generating plants was put at $37,112,000 annually. Subtracting the expected cost of the project, amortized over its fifty-year life at $33,967,000 annually, claimed power-production benefits were $3,145,000 annually. The power generated was destined for use in the large urban areas of the East and Middle West.

In addition, the power company predicted that an average of 6,230,000 people would visit the project annually for recreational purposes. Two state parks, one in Virginia and one in North Carolina, were to be donated, and the double-reservoir system was projected to become one of the most important recreational areas of the southeastern United States, with annual monetary benefits estimated to be $3,365,000. Of this figure, $2,370,000 annually was to be gained by private development induced by the influx of more than six million visitors per year.

But the true impact and meaning of the Blue Ridge Project can not be described in these statistical terms. In essence, the project would have provided electricity at times of peak demand for energy-hungry population centers far away from the New River Valley. In return, the mountain people were told to trade in their traditional lifestyle for a "modern," growth-oriented economy based on big-time recreation.

Appalachian Power Company and the FPC were offering the people of the New River Valley a new way of life. The FPC stated: "The recreation potential of the area where the project would be located is enormous. Even now there are great attractions to visitors, but when water is added, it is destined to become one of the principal recreation areas for the eastern portion of the United States." A brochure distributed by Appalachian elaborated upon this theme: "Once in existence, the upper Blue Ridge Lake will offer great potential for *economic growth*. Year 'round and seasonal homes, motels, marinas and all types of commercial and service facilities will be built around the lake. These facilities will add to the tax base of the counties as well as provide employment and increased sales. . . . Finally, Appalachian Power, itself, will pay millions of dollars of property taxes over the life of its facilities."

This new, larger project produced widespread opposition among the people of the New River Valley. Many did not consider Appalachian's new conception of life in the valley as "progress." They expressed their feelings powerfully, simply, and directly.

Blake Hampton of Ashe County wrote his congressman that "if the Blue Ridge Project is built, we will lose our 200-acre farm which has been in our family since before the Revolutionary War. I am 70 years old and my wife is 69, our next birthdays. Because of our failing health, our son, to whom we want to leave our land, is doing most of the work. We have made a good living on this land and it will produce a good living for future generations too."

Franklin D. Hubbard of Wise, Virginia, wrote to a reporter on the *Roanoke Times*, who had written an article favoring the project: "Over 40,000 acres of valuable farmland would be flooded and destroyed forever. Countless graveyards, churches and other historic homes would also be destroyed. You cry for alternatives? I'll give you an alternative. Why don't you people cut your life style and help America save some energy? . . . P.S. Flooding another man's backyard has never been right."

Brenda Sue Eddington expressed her feelings in a poem called "Decision for New River":

> Where we lived and built our homes
> And our children grew
> There all around our land
> Ran the river called New

It nurtured our land with its love
And made it pure and good
We tilled the soil and raised our crops
On places our fathers once stood

In peace we lived for many years
Until the bad news came
Men of "progress" were coming to build a plant
And our world would never be the same

The price they ask is nothing to them
Not knowing the land the way we do
For us it destroys our hopes and dreams
Along with the river called New

The people of the valley also felt that the recreational potential of the impoundments emphasized by Appalachian was an empty promise. Drawdowns necessary for both diluting pollution and generating power would eliminate swimming, since the proposed beach areas would be high and dry, particularly during the summer months. They believed that a proposed forty-four-foot vertical drawdown on the lower impoundment would leave eroded hillsides and mud flats of over half a mile. The fluctuation of the water level of the upper reservoir would, it was contended, produce mud flats of forty to seventy feet. Very difficult problems would be presented in the construction of marinas and boat-access areas, and fish would be unable to spawn under such conditions.

People also resented having to provide the water necessary to "flush" the pollution of the lower valley. They did not deem this a just or sufficient reason to abandon their ancestral homes and farms. They argued that those creating waste and sewage should have the burden and responsibility for cleaning it up and that it was unfair to make them pay for the pollution of the New downstream, three hundred miles away. The lower Kanawha was badly polluted by chemical plants and other industries, yet cities and towns were aggressively seeking new industry for the area, relying on upstream impoundments to alleviate their water-quality problems. Floyd

Crouse, a lawyer in Sparta, North Carolina, put it this way: "The people of West Virginia are asking that the isolated section of Appalachia, three hundred miles upstream from the Kanawha, where the people have made an effort to raise their standard of living by their own bootstraps, be sacrificed to a community that has an annual payroll of a hundred and sixty million dollars. It is indeed robbing a poor Peter to help a rich and affluent Paul. It is difficult to see how this could be squared with any principle of justice, equity and fair dealing, as between these two communities."

Leaders to oppose the Modified Blue Ridge Project came to the fore in the three affected counties. Floyd Crouse had been born and raised on the New River and had practiced law in Sparta all his life. He was a skilled advocate who had graduated from the University of North Carolina and the Harvard Law School, where he had been the roommate of Sam Ervin. Crouse, together with Lorne Campbell, an ardent conservationist and attorney from Grayson County, Virginia, decided to found the Upper New River Valley Association for the purpose of opposing the enlarged project. This local association was the first organized opposition in the upper valley. Crouse died of cancer on October 22, 1969. Shortly before his death, knowing of his condition, Crouse went to Sidney Gambill, a savvy lawyer who had grown up in Ashe County and practiced tax law in Pittsburgh before returning to Ashe to retire, and asked him to become president of the association. Gambill agreed and worked unceasingly against the project for the next eight years.

The FPC held new hearings on the Modified Blue Ridge Project before Judge William C. Levy in February, June, and July, 1969. The Department of the Interior, the Corps of Engineers, and the FPC staff were there to support the project. The State of North Carolina and the Commonwealth of Virginia supported the position of their citizens in the upper valley in arguing against water-quality storage for diluting downstream pollution. Governor Bob Scott of North Carolina, in a letter to the FPC, called for limiting the drawdown in the upper reservoir to ten feet so that it would be suitable for recreation. In addition, two rural electric cooperatives argued that the federal government, rather than a private company, should develop the site, and the City of Danville, Virginia, asked for a share in the ownership of the facility.

On October 1, 1969, Judge Levy handed down his Initial Decision granting the application to license the Modified Project. He included 400,000 acre-feet of storage water for diluting pollution in 1975, which was to have been the first year of operation for the project, and provided for an increase to 650,000 acre-feet by 1987. To enhance recreational use, he placed a drawdown limit on the upper reservoir of ten feet during the summer season and twelve feet during the rest of the year. The lower reservoir, it was conceded, would not have great value for recreation because of the drawdown problem. Federal development of the site was rejected, and Danville was denied an ownership interest.

Virginia and North Carolina filed exceptions to Judge Levy's decision, and the FPC held oral argument on the exceptions on February 2, 1970. The attorney general of West Virginia, Chauncy H. Browning, Jr., appeared at the oral argument and asked to intervene in the proceeding. Dramatically breaking with Governor Arch Moore of West Virginia, who supported Appalachian, Browning opposed the project because of the possible adverse impact on the warm-water fishery in the lower New River through the release of large volumes of colder water upstream. He also argued that releasing water to dilute pollution downstream was unnecessary and illegal because of the failure of the downstream chemical industries to provide adequate treatment of wastes at the source. The FPC commissioners, on April 7, 1970, granted Browning's petition to intervene, overturned Judge Levy's decision, and ordered further hearings on the project.

This was the first important reprieve gained by the opponents of Blue Ridge. Lorne Campbell and Sidney Gambill had decided, shortly after Judge Levy's Initial Decision, to try to gain the support of regional and national environmental groups. They persuaded the Washington-based Izaak Walton League to join the fight and thereby gained the services of the noted environmental lawyer Edward Berlin. The Conservation Council of Virginia, the

West Virginia Highlands Conservancy, and the West Virginia Natural Resources Council also were allowed to intervene. The Appalachian Research and Defense Fund and the Congress for Appalachian Development came in as well, represented by their able attorney, Paul J. Kaufman.

Other interests joined the opposition. The Alleghany Farm Bureau and the North Carolina Farm Bureau were allowed to intervene on November 20, 1970. Ashe and Alleghany counties, which actively opposed Blue Ridge, came to be represented by Edmund I. Adams, a lawyer from Sparta, who became, along with Campbell and Gambill, one of the chief spokesmen against the project.

Judge Levy held the next round of hearings in Beckley, West Virginia, and Washington, D.C., in July and December, 1970. The opponents of the project were present in force this time, and the sessions were stormier than ever. At that time, however, none of the opponents of Blue Ridge expected to be able to block it entirely. Even the conservation organizations stated they were not against the license but objected only to the large scope of the project and the water-quality storage component. The three states involved, North Carolina, Virginia, and West Virginia, attempted only to assure that the project to be licensed would be usable and attractive for recreation. At this point it would have been considered a victory if the opponents had succeeded in reducing the project to its original size and scope.

This round of hearings before the FPC centered around the same issues that had been discussed in 1969. More testimony was offered on both sides concerning the need for low-flow augmentation, the adequacy of the technology for treating pollution at its source, the effect on fishing in the lower New River, and the recreational benefit of the reservoirs.

Judge Levy filed his second decision, called a Supplemental Initial Decision in the bureaucratic language of the FPC, on June 21, 1971. He again concluded that the Modified Blue Ridge Project should be licensed and resolved all the disputed issues in favor of Appalachian. Two slight modifications were contained in the new decision: the downstream release rate from the lower reservoir was lessened somewhat to mollify West Virginia, and the drawdown in the upper reservoir was to be limited to ten feet throughout the year after 1985. It looked as if there was no chance to stop the project; if the FPC commissioners accepted Judge Levy's decision and gave their approval, as all expected they would, land condemnation and construction could begin.

But even before Judge Levy's second decision was announced, events were put into motion that would give a second reprieve to the New River and the opponents of the project. The National Environmental Policy Act (NEPA) had been passed by Congress and had become effective on January 1, 1970. This law required that every agency proposal for a "major federal action having a significant impact on the human environment" must be the subject of an environmental impact statement. The purpose of this requirement is to open up federal decision-making regarding the environment to make sure that agencies consider ecological, cultural, and environmental values and goals along with more traditional economic considerations in coming to any decision. The environmental impact statement must be prepared to show that the agency is aware of the environmental costs of a project and has considered them in reaching its decision.

The FPC had been slow to carry out the requirements of this new law in the proceeding to license the Modified Blue Ridge Project. Although the NEPA requirement of an environmental impact statement had gone into effect before the second round of hearings before Judge Levy in 1970, it was not until December, 1970, that the FPC ordered Appalachian to submit an "Environmental Statement" on the project. This was completed by Appalachian on January 25, 1971, but it was not until April 20, 1971, that the FPC staff submitted to Judge Levy its environmental impact statement, which was almost identical with Appalachian's. Judge Levy issued his second decision to license the project two months later, on June 21, 1971.

The environmental groups and other opponents of Blue Ridge regarded both Appalachian's and the FPC staff's treatment of environmental factors as completely inadequate. They asked the FPC commissioners to overturn Judge Levy's second decision, as they had the first, on the ground that the environmental impact statement did not accurately deal with the water-quality and drawdown problems of the project. They also charged that the impact statement had been submitted too late, a scant two months before Judge Levy's decision, and should have been prepared in advance of the second round of hearings on the project.

In addition, the project was becoming a cause célèbre in the national media. Ned Kenworthy of the *New York Times* wrote a series of stories which featured the pollution-dilution question and conveyed the idea that a rural mountain area of the country was being sacrificed for the benefit of the industrial development of Charleston, West Virginia. Bill Moyers featured the impact of the project in a documentary called "A Requiem for Mouth of Wilson" on National Education Television. Mouth of Wilson, a quiet, small town just north of the North Carolina line in Grayson County, Virginia, was to be completely inundated by the project. Through interviews with inhabitants and his commentary, Moyers painted a vivid picture of the plight of the people of the upper New River Valley.

Then, in January, 1972, while the FPC commissioners were still in the process of deciding whether to approve Judge Levy's second decision, the United States Court of Appeals handed down a decision in another FPC case* involving an application by the Power Authority of the State of New York for a permit for an electrical transmission line. The court said that the FPC had acted illegally by waiting until after the completion of hearings to prepare its environmental impact statement. It seemed obvious that this interpretation of the law meant that the FPC had committed the same error in the Blue Ridge proceeding. The opponents of the project rejoiced because it meant that Judge Levy's decision would be overturned.

But the FPC decided to fight this court decision and asked the Supreme Court to reverse the holding. The opponents of Blue Ridge waited in suspense for the Supreme Court to act, knowing

* *Greene County Planning Board* v. *FPC*, 455 F2d 412 (2d Cir. 1972).

that the license would be issued if the FPC were successful. On October 10, 1972, the Supreme Court denied review, letting the Court of Appeals' decision stand. The New River was spared for the second time, and on November 2, 1972, the FPC again refused to approve the project and remanded the Blue Ridge case to Judge Levy for a third round of hearings.

5

Another Green Light
for Blue Ridge

IF THE OPPONENTS OF BLUE RIDGE THOUGHT
that the order to hold a third round of hearings on the project
would trigger a fundamental reexamination of the situation, they
were in for a disappointment. The FPC staff, which had twice
strongly recommended the project, was in no mood to spend much
time going over it again. To the staff, it was a fundamental assumption that the nation could use all the electrical generating capacity it could get. Whenever a private power company was willing
to spend the money necessary to construct a facility, there was no
question that a license should be granted. Only the details needed
to be worked out. In addition, agency inertia and egoism came into
play. After a certain point in the planning process, a governmental
agency is totally incapable of revising its views. The Blue Ridge
case was always remanded for consideration to the very same hearing judge and staff members who had worked on it before.

As a result, scarcely two months after the remand order, on January 23, 1973, the FPC staff completed a revised Draft Environmental Impact Statement, which again recommended that Blue
Ridge be approved. The staff had consulted every step of the way
with engineers and officials of Appalachian and American Electric
Power Company, and the draft statement bore a distinct similarity
to the Environmental Statement previously prepared by Appalachian. After a few months had elapsed to allow comment on the
draft, the staff, on June 18, 1973, filed the Final Environmental Impact Statement with Judge Levy. There was no change in their view
that the project should go ahead.

On July 24 and 25, 1973, Judge Levy held hearings on the project and the impact statement as required by law. At these hearings seven FPC staff members were cross-examined by counsel for the opponents of the project. These witnesses admitted that, despite the remand for reconsideration by the FPC, the staff had not updated its evaluation of the cost of the project since the first studies were made on economic feasibility, before the Initial Decision by Judge Levy. Moreover, they had not reconsidered possible alternatives to the project; they had simply concluded that the cost of the project relative to the cost of the alternatives would be the same as before. Indeed, no new studies or information whatsoever had been gathered for the purposes of the new review. Furthermore, Judge Levy would not allow the opponents to present their own witnesses to contradict the presentation and conclusions of the staff. The new hearing and environmental statement were regarded as mere technicalities that had to be complied with before the project could be approved again.

The remand order and the new round of consideration of Blue Ridge did have an important impact in North Carolina, however. The publication of the Draft Environmental Impact Statement in January, 1973, occurred just after a new administration had taken office in that state. Governor James Holshouser, Jr., had been elected in the Nixon landslide of 1972 as the first Republican governor of the century. The new people in state government were in the mood to take a fresh look at the whole range of state policies.

Robert Finch, a young aide to Art Cooper, assistant secretary of the North Carolina Department of Natural and Economic Resources, had the largely routine job of analyzing environmental impact statements for the department. After reading the draft prepared by the FPC staff for the Blue Ridge Project, he went in to see his boss. Angered and upset, Finch pointed out the tremendous impact of the project on the two North Carolina counties, the displacement of twenty-seven hundred people from their homes, and the waste of energy involved, since the impact statement revealed that four units of electricity would be required by the project for pumping water from the lower to the upper reservoir for every three units of electricity generated. He recommended that the state try to kill the project.

Cooper, a nationally known ecologist on leave from North Carolina State University, agreed with Finch, but he knew that taking such a position would be a shift in policy for the state. Under Governor Bob Scott, North Carolina had officially opposed low-flow augmentation and excessive drawdowns but had not worked against the project itself. It was Scott's opinion that Blue Ridge would ultimately be licensed and that the only option available was to try to obtain the best possible recreational benefits for North Carolina. The state's attorney general, Robert Morgan, agreed with this view, and his assistant, Millard Rich, Jr., who had represented the state at the license hearings since 1967, had given the case low priority, merely monitoring the situation. Cooper decided the matter should be taken up with the secretary of the Department of Natural and Economic Resources, James Harrington.

Although he agreed that the project was bad for North Carolina, Harrington questioned whether the state should try to block it. He agreed to investigate further, however, and to meet with the main opponents and proponents of the project. In late February, 1973, A. Joseph Dowd, vice-president and general counsel for American Electric Power Co., presented the case for the project in Harrington's office. In a separate meeting, Edmund Adams and Lorne Campbell argued against it.

Harrington came away from these meetings with the view that Blue Ridge would disrupt the economy of the valley, cause serious social dislocations, and result in the transfer of North Carolina resources out-of-state. He felt that it was the duty of the state to try to protect its citizens, but he and Cooper agreed that if the state was going to undertake a fundamental policy shift and oppose Blue Ridge, the ultimate decision should be made by the governor. In March, 1973, Cooper and Harrington met with Holshouser and argued that the state should take a stand against the project. The governor agreed.

Holshouser's decision was primarily based on his sympathy for the people of the New River Valley. He had been raised in the mountain town of Boone, near the source of the South Fork. He knew the family-farm culture of the area. One day in the fall of 1972, when he was campaigning for governor in Ashe County, a man had approached him in tears. The man described to Holshouser what the Blue Ridge Project would do to his home and farm and pleaded with him to do something if he became governor. Holshouser had been deeply moved by the man's story, and he did not forget it. Political considerations were important as well. The two affected counties, Ashe and Alleghany, were among the few in the state with strong Republican parties, and Appalachian Power Company, since it did not do business in North Carolina, had little political clout there.

Thus the administrative apparatus of the State of North Carolina became firmly aligned with the environmentalists and the counties in opposition to the project through the decision of a Republican governor with no prior record of firm espousal of conservationist causes. Ironically, the opposite situation occurred in Virginia, where a change of administrations had also taken place as a result of the 1972 elections. The new governor, Mills Godwin, ended all state opposition to the project and firmly aligned his state on the side of Appalachian.

The first tangible expression of North Carolina's active opposition to the dam was the state's comment, written by Art Cooper, on the Draft Environmental Impact Statement. This was a wide-ranging attack on the project and on the sufficiency of the draft's analysis. Cooper accused the FPC of overstating the recreational benefits of the project while understating or ignoring the impact on fish and wildlife and the people of the New River Valley. He asked the FPC to consider other possible sites for the project or, at a minimum, a great reduction in the size of the impoundments.

This was followed by a personal letter from Governor Holshouser on July 11, 1973, to Kenneth F. Plumb, secretary of the FPC, in which the State of North Carolina's new policy was formally communicated to the agency. The governor's letter stated:

I remain deeply concerned over the social implications of this project. It appears to me that the citizens of Ashe and Alleghany Counties will bear an inordinate proportion of the negative impacts of this project and will share only distantly in its benefits. It is not at all clear to me that the recreation benefits of the project will compensate the citizens of this area for the loss of land and the disruption of way of life they will suffer. For these and other reasons, I withdraw my support for the project previously expressed by the State of North Carolina.

Thus began an unusual direct confrontation between the forces of a state government and a federal agency.

The second major consequence of the new round of consideration of Blue Ridge was the resolution, in 1973, of the long-festering controversy over the diluting of pollution. The inclusion of water-quality storage in a federally licensed water project such as Blue Ridge is a public-policy question of major significance. There are major economic and, possibly, environmental benefits for the downstream communities in regularizing streamflow so that no undue concentration of pollutants occurs during periods of natural low flow. Greater economic development becomes possible along the lower river basin, and industries and communities are spared the expense of pollution control measures at the source.

But these advantages do not come without cost. Inclusion of water-quality storage to benefit the lower basin requires a larger, more expensive project in the upper basin. In the case of Blue Ridge, the extra cost would ultimately have been borne by the customers of Appalachian and American Electric Power, since the cost of the project would have been charged back in the form of higher electric bills. In addition, in any such case the people of the upper basin themselves pay for part of the cost in the form of human suffering through the loss of their lands and homes and lost opportunity for recreation because of excessive drawdowns in the reservoirs. There also is an environmental cost, since there may be an adverse impact on fish and wildlife through regularization of streamflow, and a larger area upstream must be inundated.

In effect, low-flow augmentation grants a large economic subsidy to a downstream community. The question of the distributive justice of this subsidy comes to the fore: why should the customers of Appalachian and the people of the upper New River Valley be required to pay a subsidy to the chemical industries of Charleston, West Virginia? Why should the free market be distorted by the misallocation of resources caused by such a subsidy?

In the Blue Ridge license proceedings, these policy questions were translated into legal issues. At the time of the first two rounds of hearings, the Federal Water Pollution Control Act provided that "in the survey or planning of any reservoir by [any] Federal agency, consideration shall be given to inclusion of storage for regulation of streamflow for the purpose of water quality control, except that any such storage and water releases shall not be provided as a substitute for adequate treatment or other methods of controlling waste at the source."

The legal issue addressed during the early hearings was therefore whether it was feasible for the industries around Charleston, West Virginia, to treat their wastes adequately. Both sides offered expert testimony on this question. The Department of the Interior produced studies and expert witnesses allegedly showing that there was no possibility of at-the-source treatment. The environmentalists and the attorneys general of Virginia, West Virginia, and North Carolina offered their own witnesses, who testified that such technology was available.

Judge Levy, in each of his two decisions, found, in favor of Interior, that the level of low-flow augmentation provided would not be a substitute for at-source waste treatment and was necessary for the Kanawha River. The necessity for providing water-quality storage in Blue Ridge was the major reason for recommending a project twice as large as the one Appalachian had originally proposed.

But even as these bitter debates were being carried on before Judge Levy, a shift in public sentiment was occurring on the issue. Reflecting this, Stewart Udall, who had been Secretary of the Interior at the time that agency first intervened in the proceeding to argue in favor of low-flow augmentation, announced that he had been "misguided." As a private citizen, he wrote an article about the Blue Ridge Project, published in *Newsday* on April 24, 1971. In this article he called on the FPC to reject water-quality storage for Blue Ridge, stating that "reason dictates that the responsibility for industrial pollution must rest on the polluters, not on consumers who . . . face higher electric power rates to pay for the low-flow augmentation features of dams."

In addition, responsibility for the nation's water pollution control program had been shifted in 1970 from Interior to the newly created Environmental Protection Agency. On October 18, 1972, Congress passed the Federal Water Pollution Control Amendments of 1972. Included in this law was a new section on pollution dilution that had been introduced by Senator Sam Ervin of North Carolina with projects such as Blue Ridge in mind:

No license granted by the Federal Power Commission for a hydroelectric power project shall include storage for regulation of streamflow for the purpose of water quality control unless the Administrator [of the Environmental Protection Agency] shall recommend its inclusion. . . .

Thus the decision on the need for water-quality storage for the Blue Ridge Project was no longer up to Judge Levy. He had to abide by the decision of the Environmental Protection Agency on the matter. On April 9, 1973, the agency determined, after a review of the hearing record, that "a convincing case has not been made that the capability of providing adequate treatment at the source does not exist." It accordingly prohibited any inclusion of water-quality storage in Blue Ridge.

This should have been a clear-cut victory for the opponents of the enlarged project. The decision was legally binding on the FPC, and one would have expected the agency to return to the smaller version of the project. But this triumph, which was the basic objective the people of the upper valley had been fighting for, turned out to be an empty victory. Judge Levy, although paying lip service to the Environmental Protection Agency's decision by deleting all water-quality storage, rejected the smaller project. His third decision, on January 23, 1974, again recommended that the larger Modified Blue Ridge Project be licensed. To make up for the omission of water-quality storage, Levy simply found a need to increase the flood-control storage capacity of the reservoirs from 160,000 to

346,000 acre-feet. He also added a wholly new storage requirement of 130,000 acre-feet to improve "fishing and recreation" in the lower valley. This, he decided, was not low-flow augmentation for water-quality control but was a provision needed to regularize stream-flow in the lower basin for the benefit of *recreation*. The people of the New River Valley felt that this was a bold deception and were outraged by the decision.

6

The Scenic-River Strategy

EVEN BEFORE JUDGE LEVY'S THIRD DECISION, OF January, 1974, recommending the licensing of the Modified Blue Ridge Project, the opponents were busy devising a new strategy to save the New River. They had to find a way to circumvent the FPC, which would license the project. One day in the summer of 1973, Sidney Gambill had an idea. Congress had passed a law in 1968, the Wild and Scenic Rivers Act, that protected certain rivers from impoundment. Could the New River be declared a scenic river, he wondered, so that the FPC could not legally allow the dams to be built? In 1971 the FPC had been compelled to place a four-year moratorium on the construction of a hydroelectric project on the Hell's Canyon section of the Snake River in Idaho in order to allow Congress time to consider whether the river should be included in the National Wild and Scenic Rivers System. Perhaps this tactic would work for the New River.

In order to get a moratorium on Blue Ridge, Congress would have to act. In the fall of 1973, Senator Sam Ervin of North Carolina and Congressman Wilmer Mizell, whose district included the upper New River Valley, introduced identical bills in the Senate and the House of Representatives to require the Department of the Interior, which administered the Wild and Scenic Rivers System, to undertake a study of the New River to determine whether it was suitable for designation as a national scenic river. This proposed law, if passed, would give the river affirmative protection against the FPC license during the period of the study and could block the dams altogether if, at the end of the study period, Congress should pass another act making it a component of the Wild and Scenic Rivers

System. These legislative initiatives came too late to be passed before the end of 1973, however. Congress adjourned without acting on the matter.

After Judge Levy announced his decision in January, 1974, action was taken in North Carolina to implement the scenic-river strategy. On January 24 bills were simultaneously introduced in both houses of the state General Assembly to designate part of the New River as a North Carolina scenic river. Hamilton Horton, a North Carolina state senator from Winston-Salem, was the primary force behind this drive. Unlike federal scenic-river designation, naming the New River in the state system would not block the project, since the FPC had the power to override an inconsistent state law. State designation could, however, aid the effort being made in Congress and would provide an additional factor for the FPC commissioners to consider in deciding whether or not to follow Judge Levy's recommendation.

On March 21, 1974, these efforts met with success. The General Assembly of North Carolina formally included 4.5 miles of the river from the confluence of the North and South forks north to the Virginia state line as a state scenic river. In addition, a resolution was passed calling on the North Carolina Department of Natural and Economic Resources to study the South Fork of the New River to determine the feasibility of declaring it a state scenic river. The speed with which the General Assembly completed action on these measures astonished even their proponents.

This action on the state level was accompanied by another attempt in Congress. Senator Ervin, joined now by North Carolina's other senator, Jesse Helms, again introduced a bill in the Senate to study the New as a possible national scenic river. This measure proposed to amend section 5 of the Federal Wild and Scenic Rivers Act to add the New River in North Carolina as "a potential addition to the Wild and Scenic Rivers System." The Department of the Interior would conduct a feasibility study and report back to Congress. A separate law would then be needed to designate the New as a scenic river, thus ultimately blocking the project.

The two-year study bill was designed to buy a little more time for the river.

On February 7, 1974, the Public Lands Subcommittee of the Senate Interior Committee held hearings on the bill. A. Joseph Dowd, general counsel for American Electric Power Company, testified in opposition to the bill, calling it a "last-ditch attempt to kill Blue Ridge." He called on the committee to leave the matter in the hands of the FPC. Senator Ervin, on the other hand, argued that the river deserved to be studied for inclusion in the scenic-rivers system before it was destroyed by the power project. Edmund Adams testified that the people of the valley just wanted to be left alone, and he asked for favorable consideration. The Department of the Interior, which would have to undertake the study of the river, was ambivalent about the bill. At the February 7 hearing, A. Heaton Underhill, assistant director of Interior's Bureau of Outdoor Recreation, asked that the bill be deferred. Interior wanted a study of the upper forks of the New River outside the project area, which would not interfere with Blue Ridge.

In March, 1974, Governor Holshouser took matters into his own hands to break this impasse. He held a personal meeting with Rogers Morton, the Secretary of the Interior, to persuade Interior to adopt North Carolina's position. Morton, who was later to serve as President Ford's campaign chairman for the presidential primary contests against Ronald Reagan, was very much a political animal. It was not lost on him that Holshouser was a key Southern Republican governor whose support against Reagan could be crucial.

As a result of the meeting, Morton swung Interior four-square behind North Carolina's effort to delay Blue Ridge. In a letter to Senator Helms on April 4, 1974, Morton wrote:

I share your belief that the development of adequate energy supplies is a critical national priority; but I believe as you do, that we must weigh carefully the environmental impacts of such development. I am not convinced at this time, without benefit of further study, that it would be in the interest of sound resource management to forego 40,000 acres of land and 70 miles of heretofore free-flowing river for the sake of pumped storage peaking capacity which exceeds projected demand.

On April 5, 1974, the Public Lands Subcommittee, in possession of Morton's letter, unanimously approved the New River bill and reported it to the full Senate Interior Committee. On May 2 that committee unanimously adopted the bill by voice vote and reported it favorably to the Senate.

The measure was considered by the full Senate on May 28, 1974. The debate was sharp and, at times, bitter. Senator Haskell of Wyoming offered an amendment to the bill limiting the time of the study to two years. The North Carolina forces agreed to this, and the amendment was adopted. Senator Ervin opened debate on the bill, declaring that the New River was a worthy addition to the Wild and Scenic Rivers System:

I am not an expert in the technology of producing electric power and, therefore, must leave the resolution of this particular controversy to others. But one thing is certain. . . . Whether this project be technologically sound or not, its completion will destroy the New River. And . . . with respect to the New River and the joy it brings to those who know it, I do qualify as an expert. I am an expert as to its beauty because, like most other fellow travelers on this Earth, I have learned to appreciate the handiwork of Almighty God.

Ervin argued that he was not asking the Senate to disagree with the FPC as to that agency's possible determination that the license should be issued. Rather, the problem was that the FPC knew only about electric power production; it did not have the competence to judge whether the river should be declared a national scenic river. Thus, if the bill failed to pass, this alternative use of the resource would not have been considered and balanced against power production. In closing, Senator Ervin quoted some letters he had received from schoolchildren in Alleghany County. One poem read, "This is what I want. Senator Sam—stop the dam."

Senator Helms spoke next, directly attacking the FPC's anticipated decision on Blue Ridge:

The studies which the FPC has made are based on old concepts of resource management. . . .

The pumped-storage concept of supplying peaking power has been
shown to be an uneconomic concept when considered in the overall
context. Steam-generating plants will be used to pump water up to the
higher reservoir at night so that it can flow down through the water tur-
bines during the day when the peak demand is highest. But there is no
such thing as perpetual motion. It will take three units of power to
generate two units produced by the pumped-storage facility. . . .

It is highly significant that this project has a planned drop of yearly
output of nearly two-thirds over 20 years, from 5.5 to 2.75 million mega-
watt hours. The reason for this is that the reservoir will silt up at a rapid
rate, leaving no room for the quantities of water needed for economical
generation.

The forces opposed to the passage of the scenic-river study bill
were led by the two senators from Virginia, Harry F. Byrd and
William L. Scott. They argued that the dispute was basically between two states, North Carolina and Virginia, and that Virginia
would get substantial benefits and therefore favored Blue Ridge.
They called the study bill an ill-conceived attempt to stop Blue
Ridge. Senator Scott summed up the Virginia position:

It is not reasonable to make this project a political football to see
which State has the greater political muscle. Neither Virginia Senator
is asking the Senate to approve the project. We are merely asking that
you leave the decision of whether a private utility company can construct a hydroelectric dam with private capital to the Federal Power
Commission. That is what the Commission was created to do. We do
feel, however, that a State with only 21 miles of a 254-mile stream has a
far less interest in a project than the adjoining States of Virginia and
West Virginia with the remainder of the river and that the views of the
State of Virginia of where the project is to be constructed should not be
ignored. The defeat of this proposal and a favorable decision by the
Federal Power Commission will provide new clean energy without cost
to the Government at a time when the American people are concerned
about both energy and our environment. . . .

To me, S. 2439 [the scenic-river study bill] constitutes a misuse of

National Wild and Scenic Rivers legislation. Its purpose is to block a badly needed power project. . . .

After more than 9 years of exhaustive proceedings during which everyone has been permitted to have his say, the Federal Power Commission is about to render a decision on the merits of Blue Ridge. FPC's decision will be based upon a comprehensive evidentiary record covering all aspects of the project: power, recreation, flood control values, social and environmental impact; and the full range of alternatives to Blue Ridge—including the preservation of the river in its existing state. . . .

The Congress does not have all of these facts before it. We, therefore, urge that FPC be permitted to reach a decision on the merits of Blue Ridge.

When the Senate voted, the result was a lopsided victory for North Carolina and the scenic-river study bill. By a vote of forty-nine to nineteen, the bill passed the Senate.

Attention now turned to the reaction of the FPC to the Senate's vote. Opponents of Blue Ridge hoped that this May 28, 1974, vote would once again cause the FPC commissioners to delay a decision on the license. It soon became evident that this would not be the case. Rumors reached North Carolina officials in early June that the FPC was going ahead. Edmund Adams, on behalf of Ashe and Alleghany counties, sent a telegram to the commission requesting the opportunity to present additional arguments to it before the license was issued. Several members of Congress also wrote to the FPC asking that a decision on Blue Ridge be delayed.

But the FPC had decided to act. It felt that its jurisdiction and authority were imperiled, and it was not about to give in on Blue Ridge as it had on the Hell's Canyon dam project. The commission realized, however, that it could not license the project without making at least some concession to the effort being made in Congress to study the river for inclusion in the Wild and Scenic Rivers System. Its action would be reviewable in the courts, and it could not appear to be in defiance of this legislative initiative which had already passed the United States Senate.

The commission solved this problem by resorting to a lawyer's tactic. It decided to issue the license immediately but not to make it effective until January 2, 1975, thus giving Congress a deadline of seven months in which to act. If the scenic-river study bill passed within this time limit, the project would be delayed; if Congress did not act, the license would automatically take effect, and Blue Ridge would go ahead.

Thus, on June 14, 1974, the FPC formally issued a license for the Modified Blue Ridge Project to the Appalachian Power Company. The commission cited the many years of consideration of the project and found that it met the requirements of section 10(a) of the Federal Power Act in being the alternative best suited to a comprehensive plan for "improving and developing" the New River waterway. The commission accepted Judge Levy's recommended decision, including the elimination of water-quality storage and the limitation of the drawdown on the upper reservoir to a maximum of ten feet. The commission slightly decreased the size of the double impoundments, from 44,000 to 42,000 acres, by eliminating 186,000 acre-feet from the total of 346,000 acre-feet recommended by Judge Levy for flood control. But the commission accepted Judge Levy's recommendation for 130,000 acre-feet to augment natural flow in the lower basin to improve fishing and downstream recreation in West Virginia.

Three major benefits were claimed for the project. First, peak-demand power capacity would be provided for use in the American Electric Power Company's service area in the east-central states. Second, enhanced recreational opportunities would result from the donation by Appalachian of land for two state parks, one in Virginia and the other in North Carolina. The commission predicted that the area would become one of the principal recreation areas of the entire eastern United States. Third, flood control would be a major purpose of the project. The flood-control storage would protect the Valley of Virginia sufficiently to reduce property damage by 72 per cent, calculated on the basis of the highest flood on record, which occurred in 1940.

As for the scenic-river option, the commission's stand was that it had considered and rejected the option of allowing the river to remain in its natural condition:

As for the elimination of a free-flowing river, and the substitution for it of sizeable, flat lakes, we cannot do otherwise than share in the regret. We do not authorize the project because our regard for rivers and streams is incomplete, but because we find it necessary to balance that regard with the need for the benefits that the project would create.

Despite the FPC's decision to grant the license for the Blue Ridge Project, North Carolina and the opponents remained optimistic that they could once again get a reprieve for the New River. The scenic-river study bill was proceeding rapidly through the House of Representatives. On June 3, 1974, even before the FPC's license was issued, the National Parks Subcommittee of the House Interior Committee had held hearings on the bill. On August 21, 1974, the Interior Committee reported the bill favorably by a vote of twenty-one to fifteen. The bill then went to the House Rules Committee, which would have to vote in favor of the measure before it could be brought to the floor of the House. It looked as if the seven-month deadline would easily be met.

During the fall of 1974, the members of the House Rules Committee came under intense lobbying pressures from both proponents and opponents of the project. The power-company forces decided that focusing on the Rules Committee would provide their best chance to block the bill. The electric-utility industry regarded Blue Ridge as a test case, and A. Joseph Dowd, who personally headed their lobbying forces, argued that if the project were further delayed, private utility companies generally would be reluctant to go through the expensive FPC licensing process in the future. The Sprague Electric Company, which operates a plant in the project area in Ashe County, stated to committee members that blocking Blue Ridge would have a chilling effect on that county's industrial potential.

The entire Virginia congressional delegation, Virginia Governor Godwin, and Virginia Attorney General Andrew P. Miller also lobbied against the bill, citing the benefits Blue Ridge would have for Virginia and the fact that it could not be built without use of the North Carolina section of the New River.

The most potent lobbying forces against the bill, however, were those of the AFL-CIO and of organized labor. George Meany, the head of the AFL-CIO, looked upon Blue Ridge as a source of new jobs in the construction industry, which was severely depressed in 1974. Many of the Rules Committee members had close labor ties and were subject to labor pressures.

On the other side of the controversy were the entire North Carolina congressional delegation and the North Carolina state government. They were joined by officials of the Department of the Interior and by Congressman Ken Hechler of West Virginia, who wished to include the New in that state also in the Wild and Scenic Rivers System.

In addition to the lobbying activity, the scenic-river study bill was delayed by the chairman of the Rules Committee, Ray J. Madden of Indiana. Madden refused to permit a vote until the Senate Interior Committee reported favorably an unrelated bill to enlarge the Indiana Dunes National Park on the southern shore of Lake Michigan. In effect, one piece of environmental legislation was being held hostage to obtain passage of another. This critical delay allowed the proponents of Blue Ridge additional time to work on Rules Committee members.

This waiting game was played into December, 1974. It began to appear doubtful whether the Rules Committee would vote at all on the matter or, if it did, whether the full House would have time to consider it before the impending adjournment for Christmas.

On December 9 Evelyn Holland, who had grown up in the New River Valley and was now anxiously following the outcome of action in Congress, wrote a touching letter to the editor of the Raleigh *News and Observer*, which expressed the mood of many who knew and loved the New River:

Steve Berg's article . . . in Sunday's edition . . . brought a flood of memories and tears over the impending death of a beautiful friend.

Being reared in Ashe County near New River, in youth I experienced the rugged life of the mountaineer family struggling for a living. There were lean years, cold winters, hours spent tending the farm and livestock, long bus rides over bumpy, rutted dirt roads to school. Until I was 14 there was no electricity. Telephones were installed after I graduated in

1955 and left for college. The winding road was hard-surfaced more recently. Homes with running water were few and far between at that time. We had to keep logs for the heater and split wood for the cook stove. We were financially poor and so were our neighbors.

My background sounds like a bad dream, you think. The truth of the matter is that memories of my childhood in Ashe County near New River are most precious. We knew we were poor, so we learned thrift; we never lacked necessities, were hardy, independent and creative. "Poverty stricken" was a term we had never heard.

What we did have was peace. By living close to nature, we experienced the splendors of the seasons from our doorstep. The cold and icy winter melted in the warm spring sun and new life painted the land with a different set of fresh new colors and filled the air with pleasant sounds and smells. The fullness of summer was the rich taste of fruits and vegetables we had grown from seed, and butterflies and birds competing with wildflowers for brilliance in the sun, and leaves so dark green they were almost black. Autumn meant more colors and the last of the food cared for and added to the supply which would see us through another winter and leave plenty for sharing.

Fresh water from cool springs seeped between rocks and formed little branches which trickled down the hills and emptied into rushing creeks that flowed to the river.

This was ours. We were close to nature. We were close to God. We were rich!

Because of a proposed dam, the river could die and much of Ashe, the fairest of our 100 counties, would be flooded by a lake which will be surrounded much of the time by a mosquito infested, stinking mud flat. Land developers, resort homes, motor boats, pollution, plastic, concrete, neon signs and "progress" will finally come to Ashe County.

Finally, a week and a half before Congress was to adjourn, Madden relented under pressure from both the North Carolina delegation and from his own constituents, who had been enlisted on the New River's behalf. The scenic-river study bill was put on the agenda for Wednesday, December 11.

But it was too late. When the vote was taken, the Rules Committee, by a vote of thirteen to two, denied the bill the opportunity to come to the floor of the House. The backers of the bill had simply been outgunned by the combined forces of the utility industry, organized labor, and the State of Virginia.

The North Carolina forces refused to concede defeat, however. There was one last hope. Under the rules of the House, it was possible, if the House leadership agreed, to allow the full House to vote on whether to suspend the rules in order to pass the scenic-river study bill. Although this parliamentary maneuver would circumvent the Rules Committee and allow the bill to come to the floor, the measure required a two-thirds vote to pass.

The Speaker of the House, Carl Albert, who was willing to consider the matter, held conferences with both the Virginia and North Carolina delegations. The North Carolinians, headed by Congressman Roy Taylor, the second-ranking member of the House Interior Committee, argued that allowing a full House vote was the only fair way of handling the matter and that its fate should not be decided by a fifteen-member committee. Governor Holshouser sent a personal telegram to Speaker Albert asking for a suspension of the rules and a House vote.

On Monday, December 16, Albert agreed to bring the matter before the House. He limited debate to forty minutes and scheduled a vote for Wednesday, December 18. On Tuesday James Harrington, secretary of the North Carolina Department of Natural and Economic Resources, flew to Washington and spent the day personally lobbying for the bill. Governor Holshouser sent telegrams to all 435 House members.

The debate on the New River was held on the House floor on Tuesday evening, December 17. Congessmen Taylor and Mizell questioned the need for Blue Ridge and spoke of the great human costs of allowing the license to become effective. Mizell, in addition, attacked the FPC as an "uncaring bureaucracy" and the "power monopoly of the power utilities and labor unions." Both cited the fact that Rogers Morton, Secretary of the Interior and the newly appointed chairman of the President's Energy Resources Council, opposed Blue Ridge. Ken Hechler of West Virginia spoke poetically, calling the New "a great river which cascades through gorges and canyons, transecting every ridge of the Allegheny Mountains, and roaring through some of the wildest whitewater of the East." He compared the idea of damming it to "dynamiting the pyramids" and said the project would create work only in the sense that "driving yourself into a telephone pole makes jobs for fender repairmen."

William C. Wampler of Virginia, whose district included Grayson County, the site of the dams, as well as Roanoke, the headquarters city of Appalachian Power Company, called on the House to reject the bill and not to overrule the considered judgment of the FPC. Congressman Daniel of Michigan, reflecting the view of labor, called the project a boost to the economy, since it would result in $500 million of private capital investment, creating jobs for twelve-hundred people, with an annual payroll of $120 million for six to eight years. Sam Steiger of Arizona derisively referred to the bill as a block-the-dam bill, not a scenic-river study bill.

Morris Udall of Arizona eloquently ended the debate by calling the attention of the House to the fact that the bill seemed to be the last chance for the New River: "There are decisions which are irrevocable and if tonight we fail to pass the bill by two-thirds, the bell has tolled on the New River. It is gone. The jobs are gone. The farms are gone. The land is gone."

On the next day, December 18, the question was called and the vote was taken. One hundred ninety-six voted in favor and 181 were against, with 57 not voting. Although a majority favored passage, support fell far short of the necessary two-thirds. The Rules Committee's negative vote had prevailed after all, and those in favor of Blue Ridge had won. The New River had lost, and Congress adjourned. The new Congress would not convene until January, 1975, after the license had taken effect.

7

Counterattack in Court

WITH THE DEFEAT OF THE SCENIC-RIVER STRAT-
egy by Congress in December, 1974, it looked as if the license for
Blue Ridge would become effective, as scheduled, on January 2,
1975. Appalachian would then be free to proceed with condemnation of land and construction of the dams. For the opponents of the
dams, it was crucial to postpone the effective date. There was only
one place to look for such quick action—to the courts.

The procedure for going to court to appeal an FPC license is
quite complicated. Before invoking the aid of the court, the person
contesting the license must file what is called a "petition for rehearing" with the FPC. This is a legal document formally requesting that
the FPC reconsider the matter. Only after the FPC formally denies
the petition for rehearing can the case be taken to court. Furthermore, only those arguments that are raised in the petition for rehearing can be raised in the subsequent court action.

In the summer of 1974, after the FPC's decision to grant the
license was announced on June 14, the opponents of Blue Ridge
complied with this procedure; but it was done badly, since everyone's attention was focused on the scenic-river strategy in Congress.
In the first place, a mix-up occurred concerning the date on which
the petition was due. The thirty-day deadline was July 14, but since
this was a Sunday, an extra day was allowed, until Monday, July 15.
Under FPC rules, a document is filed on time only if it is *received*
at the FPC in Washington by the due date. The lawyers for the opponents of Blue Ridge were unaware of this rule, and they took the
thirty-day deadline to mean that the petition would be on time if
it was *postmarked* on the due date. Accordingly, the FPC rejected
five of the six petitions filed because they had not been received

by July 15. These included the petitions of all the local opponents of Blue Ridge as well as all the conservation organizations: Ashe, Alleghany, and Grayson counties, the Congress for Appalachian Development, the Appalachian Research and Defense Fund, the New River Chapter of the Izaak Walton League, the Upper New River Valley Association, the New River Pioneer Chapter of the Daughters of the American Revolution, Young's Chapel Baptist Church, and Sidney B. Gambill. The possibility of court review would have been cut off completely had it not been that one party, the State of North Carolina, filed on time. This occurred largely by accident, as the lawyer who prepared the petition for North Carolina, knowing he would be out of town on Monday, July 15, prepared and mailed the petition on the Friday before.

A second problem, which would loom larger as time went on, was that the on-time petition, North Carolina's, had been hastily done and raised only general objections to the license: that the FPC was in error in concluding that the project would contribute more than it would take away and that the agency had not followed the procedures required by law. On August 12, 1974, the FPC denied North Carolina's petition for rehearing and wrote a thorough, twenty-six-page refutation of all the challenges raised. It pointed out that all of North Carolina's earlier specific objections to Blue Ridge had been removed. Water-quality storage had been completely eliminated, and the drawdown problem had been lessened since it was now limited to ten feet in the upper reservoir. Appalachian had been required to donate land to North Carolina for a state park. The commission cited the huge hearing record of more than forty thousand pages of testimony compiled over a twelve-year period as evidence that it had thoroughly considered every aspect of the project. North Carolina's latest general objections to the project were disposed of by an equally general rebuttal referring to the "beneficial environmental consequences" of Blue Ridge.

Since the FPC had denied North Carolina's petition for rehearing, the state was free to file an appeal in court. A notice of appeal was formally filed in the United States Court of Appeals in Washington. The case was before the court, but the grounds for appeal being relied upon by the state had been convincingly refuted by the FPC. At that time, however, this seemed unimportant, since in the fall of 1974 the scenic-river strategy seemed to be working in Congress, and it looked like a waste of time to prepare for a costly court battle.

It was at this point that the author became involved in the proceedings. One day in late August, 1974, I received a phone call at my office from Ernie Carl, an environmental planner for the State of North Carolina. I had met Ernie when he was a zoology professor and a colleague at the University of North Carolina. Now he was working in state government, and as the close aide of James Harrington, he had been given the job of coordinating North Carolina's political effort to save the New River. He knew I taught environmental law at the university, and he asked me to work informally with the attorney general's office on the New River appeal. I readily agreed, although we talked about the fact that Congress would probably act first and that there was little possibility we would actually go to court.

At first I gave the matter low priority. It seemed as if the House Rules Committee would report the scenic-river study bill for a full House vote. There we had the votes to win, and Governor Holshouser was confident the President would sign the bill. As the days went by, however, I grew increasingly worried. By the time we knew whether Congress would pass the scenic-river study bill, it would be December, just before adjournment, too late to formulate any new strategy to save the New River. Something should be done as a safeguard against the failure of Congress to act. Yet there were no plans for such a contingency. Furthermore, I felt that the grounds raised by North Carolina in the petition for rehearing, to which the state would be limited on appeal, were not legally sufficient to challenge the FPC's decision in court.

I spent most of September puzzling over this. It seemed that the situation held a certain cruel irony. In most environmental cases, relatively powerless and impecunious conservation organizations are up against the vastly superior forces of the state and federal governments. In the New River case, we had the State of North

Carolina on our side, but it seemed that nothing could be done if Congress did not act.

Slowly, a strategy formed in my mind. First of all, we had to build up some new intellectual capital and formulate viable legal theories to challenge the license in court. I decided that we had to develop the best legal arguments we could think of without regard to whether they were raised in the petition for rehearing. We had no choice but to risk being told by the court that we had no right to raise the arguments. After all, opposing counsel might not object to our new arguments, or we might successfully argue that the general grounds raised in the petition were legally sufficient to allow us to use the more particular new arguments. We were groping for straws. I decided that a draft brief should be prepared on the assumption that we might have to go to the Court of Appeals on short notice. Janet Mason, then a third-year law student at the University of North Carolina, agreed to help write it.

By the end of October, we had completed the brief. It raised three basic arguments against the FPC's decision to license Blue Ridge:

I. *The FPC violated the National Environmental Policy Act (NEPA) in licensing Blue Ridge by refusing to allow consideration of the New River as a national scenic river.*

One of the procedures the NEPA requires when a resource decision is made by a federal agency is that, before action is taken, all alternative uses or dispositions of the resource be analyzed and studied. The purpose of this requirement is to ensure that a decision be based on adequate information and the balancing of competing societal values and goals. In the case of major action such as Blue Ridge, the agency must document its consideration of alternatives in writing as part of the environmental impact statement. This is to allow full disclosure to all interested parties: the Congress, federal and state agencies, and the public. The FPC's own regulations require this. Environmental impact statements must "fully deal with alternative courses of action to the proposal and, to the maximum extent practicable, the environmental effects of each alternative."

In the case of the Blue Ridge licensing proceeding, it was obvious, at least after 1973, that the most important possible alternative use for the New River was its inclusion in the National Wild and Scenic Rivers System. The Environmental Protection Agency, in its comment on the Draft Environmental Impact Statement, had asked the FPC specifically to address the features of the river that might warrant its designation as a national scenic river. Despite this, in the Final Environmental Impact Statement the FPC never studied or discussed this alternative beyond merely stating that failure to license Blue Ridge would preserve the river for possible later selection as a wild or scenic river.

The FPC not only failed to consider the scenic-river alternative for the New River but also tried actively to cut off study of this possibility by other responsible branches and agencies of government. Judge Levy issued his third decision recommending the licensing of Blue Ridge on January 23, 1974, in spite of the fact that Congressman Mizell had specifically called to his attention the fact that the scenic-river study bill had been introduced in Congress. Later that year, the full FPC refused to defer action on Blue Ridge and went ahead and granted the license on June 14, 1974, at a time when the Senate had passed the scenic-river study bill by a lopsided vote and hearings had been held in the House. The FPC's only concession was to delay the effective date of the license to January 2, 1975, thus arrogantly giving Congress a deadline by which to act and attempting to subvert unbiased consideration of the measure. The cumulative effect of this conduct was a gross violation of the NEPA and a perversion of the FPC's responsibility to consider the broad public interest rather than narrow energy-supply concerns.

II.   *The FPC, in licensing the Blue Ridge Project, violated the National Environmental Policy Act (NEPA) by failing to consider conservation of energy resources as an alternative to the project.*

The FPC, in licensing Blue Ridge, took the position that it did not have any duty to look at whether energy conservation would affect the need for the facility. The environmental impact statement dismissed energy conservation out of hand: "Even if all U.S.

consumers were to adopt stringent conservation practices with respect to electricity, the demand for electric power would continue to increase to such an extent that all of the production of Blue Ridge would be essential and, in fact, would meet only a small part of the foreseen future need." Citing pre-energy-crisis growth predictions in its 1970 *National Power Survey,* the FPC stated, "The electric load of [American Electric Power Company] has been growing at a fairly steady rate and it is anticipated that it would continue to do so."

But the NEPA requires that an agency study and consider all alternatives. This consideration should not be limited to alternative methods of power generation but should also include possible measures for reducing consumer demand that would reduce the need for a facility the size of Blue Ridge. This was especially important in the case of Blue Ridge, which, as a pumped-storage facility, would consume four units of electricity for every three produced.

The reason for having to license and build plants such as Blue Ridge is that, until recently, there has been no attempt by state utilities commissions or the FPC, which have the authority to approve rates, to charge more for power consumed at peak periods and thus no consumer incentive to avoid such consumption. As a result, peak demand has grown immensely, at about one per cent a year faster than average demand for electricity. The annual load factor (the ratio of average-load consumption to peak-load consumption) of utilities in the United States declined from 66 per cent in 1960 to an estimated 61 per cent in 1975. This means that power companies, on an average basis, have idle generating capacity of 39 per cent. Yet added investment and new capacity have been constantly required, since FPC and utility-industry policy mandates new facilities adequate to take care of about 17 per cent over and above the highest projected peak demand.

Under such a system, pumped-storage projects such as Blue Ridge are considered the solution for the peak-demand problem. They provide a source of instant-on, peak-demand energy that can be tapped more efficiently than that of steam electric plants, which cannot be brought on or taken off line so easily. In addition to satisfying peak demand more efficiently, the utilities' load factor is improved, since pumped-storage plants consume great amounts of electricity during off-peak hours, requiring existing steam plants to operate on a more continuous basis. Thus it is of no consequence to the utility that four units of power are consumed for every three that are produced.

But on closer analysis, this is an unacceptable solution. The load factor is improved, not by any real savings or efficiency, but because through this system *the utility itself becomes a main consumer of its own product*. This also ensures the need for additional expansion of the system. The result works to the utilities' financial advantage, since under the law their profits are set by state agencies and the FPC based upon the size of their investment and costs plus an allowed rate of return.

Society and the public, however, cannot afford to keep up with peak demand through allowing unrestrained new investment. In the past technological improvements generally meant that when expansion took place, the utilities reduced their unit or average costs. Today, newly installed capacity is being brought on line at significantly higher costs per kilowatt than historical costs.

A better way to attack the peak-demand problem and to help eliminate the need for facilities such as Blue Ridge is for the FPC and state utilities commissions to compel the introduction of peak-load pricing, time-differentiated rates for electricity. This would discourage consumption at peak hours by higher charges for service when the demand is greatest, thus providing an incentive for more use of off-peak capacity and load-factor improvement. Time-differentiated rates make sense because they are based on marginal-cost economic theory rather than, as in the present system, on fully distributed historical costs. Customers are charged in proportion to the unit costs they impose on the system, and peak-demand customers pay more because it costs more to serve them. Customers are allowed the freedom to choose the consumption pattern they can best afford. Since 1975, state utilities and public service commissions all over the country have begun to study and implement peak-load pricing.

Despite this interest and the fact that the FPC approves interstate wholesale electric rates and could itself help implement peak-load pricing, this alternative was not studied prior to licensing Blue Ridge. Other possible technologies for storage of electricity for later use were not considered, and even other alternative pumped-storage sites with a cheaper capital cost per kilowatt than Blue Ridge were ignored. In basing its licensing decision on 1970 demand projections, the FPC also disregarded possible slower growth in the consumption of electricity and did not consider possible conservation measures to further slow overall growth.

III. *The FPC's environmental impact statement did not adequately disclose the true costs and benefits of the Blue Ridge Project.*

Construction costs for the Blue Ridge Project were estimated to be $430 million, and claimed net annual power benefits, calculated on the basis of cost savings of the project compared with coal-fired plants, were $3,145,000. Net annual recreation benefits were claimed in the amount of $3,365,000, a figure larger than the net annual power benefits. Since the FPC admitted that other pumped-storage sites could provide peak-demand power at lower cost, the recreational benefits provided the crucial margin necessary to justify the project. But those benefits were admitted to be "speculative" by the FPC, and they were questioned by numerous witnesses throughout the hearings on Blue Ridge. It was pointed out that the lower reservoir, with its forty-two-foot fluctuations, would have little recreational potential and that the upper reservoir would be subject to ten-foot vertical drawdowns, producing forty to sixty feet of mud slopes. The pumping cycle would place the low point in the cycle on Friday evenings, when most visitors would arrive. There were also other reservoirs nearby, not subject to pumped-storage drawdowns, that could be used for flat-water recreation. The claimed benefits for fish and wildlife overlooked the fact that the project would inundate a large portion of the oldest river in the nation, an unpolluted, free-flowing stream ideal for canoe recreation, and would destroy a great amount of fish and wildlife habitat.

Flood-control benefits were also included without any indication of how they were arrived at or whether flood protection could be provided by other measures, such as land-use restrictions in the floodplain. The FPC cost-benefit analysis also failed to disclose the value of agricultural sales in the flooded area that would be taken away. These were estimated to amount to $13,500,000 annually by agricultural extension offices in the affected counties.

The FPC impact statement also claimed that Blue Ridge would mean less air pollution because the project would allow coal-fired plants in the system to operate more efficiently. This contention was disputed by the Environmental Protection Agency, which pointed out that Blue Ridge would require existing generating units to operate for longer periods of time and that these units were subject to much less stringent sulfur-emission standards than new plants, so that sulfur dioxide emissions would increase.

The most serious failing of the Blue Ridge impact statement, however, was the lack of any attempt to deal with the social impacts and costs of the project. These were even more important than the adverse environmental impacts in the affected area. The disruption of an established pattern of life dependent on the bottomlands of the river and its tributaries was not mentioned, let alone studied. Human dislocation was treated only in economic terms. The FPC never held hearings in any of the affected counties.

The social impact of the project would be tremendous and would involve much more than the great hardship of the twenty-seven hundred to twenty-eight hundred persons who would be directly displaced. Many other small farmers would not lose their homes but would be deprived of their most productive agricultural lands. Other families would lose access to churches, schools, and commercial facilities. A boom-and-bust economy would be created. The boom would occur during construction of the project, but in the long run it was uncertain whether the reservoir recreation would have any appreciable impact on local employment and personal income. A study of Norris Lake, a Tennessee Valley Authority reservoir, by Professor Charles Garrison, an economist at the University of Tennessee, concluded that "recreation's contribution to the local economy has been negligible." The same might be true for Blue Ridge.

While we were working on this brief, it struck me that we ought to do something more to prepare for the contingency that Congress would not act. I knew that we would not be able to accomplish anything by our present approach until the case came up on the regular docket of the Court of Appeals. We had to be ready to get a court order delaying the January 2 effective date of the license. Under the Federal Rules of Appellate Procedure, it is possible to get such an order; the court can stop a project temporarily—in legal terminology the project is *stayed*—in order that there may be time for full judicial review.

I decided to develop a strategy for obtaining such a stay by the Court of Appeals before the January 2 effective date. I had gotten an idea from a case called *Warm Springs Task Force* v. *Gribble*, which had just been decided in the United States Supreme Court. In that case Justice William O. Douglas had granted a stay of the construction of a dam in California. His chief ground for doing so was a letter from the general counsel of the President's Council on Environmental Quality (CEQ) to the solicitor general of the United States in which the CEQ raised objections as to the adequacy of the environmental impact statement. In the course of his opinion, Douglas stated:

> The Council on Environmental Quality, ultimately responsible for administration of the NEPA and most familiar with its requirements for Environmental Impact Statements, has taken the unequivocal position that the statement in this case is deficient, despite the contrary conclusions of the District Court. That agency determination is entitled to great weight . . . and it leads me to grant the requested stay. . . .

It would help immensely the New River's chances of getting a court to stay the January 2 effective date, I believed, if a similar letter could be obtained from the CEQ stating that the FPC's impact statement for Blue Ridge was inadequate. I called Ernie Carl and suggested that the CEQ might respond if Governor Holshouser personally wrote a request to Russell W. Peterson, who was the chairman of the CEQ and had previously been a Republican governor of Delaware.

One of Carl's former students, Bob Smythe, was, by coincidence, working at the CEQ and was in charge of overseeing federal water-resources projects. Carl called Smythe to alert him that he was about to write, for Governor Holshouser's signature, a letter to Peterson asking for a formal CEQ review of the Blue Ridge impact statement. Smythe answered that he was generally familiar with the project but that the CEQ ordinarily did not like to comment once a case was in litigation; he would see what he could do, however.

When Peterson received the letter from Governor Holshouser, he referred it to his staff director, Steven Jellinek, who, in turn, assigned it to Smythe. Smythe analyzed the impact statement for Blue Ridge and wrote a letter of comment for Jellinek's signature. This letter was sent to Kenneth Plumb, secretary of the FPC, on October 30, 1974, with a copy to Governor Holshouser.

Pursuant to a request from the Governor of North Carolina, the Council has reviewed the Commission's final environmental impact statement (EIS) for the Modified Blue Ridge Power Project, No. 2317, dated June 1973.

We are aware of the pending legislation to authorize a study of the New River for possible inclusion in the National Wild and Scenic Rivers System, and we commend the Commission's action to delay granting of a license for this project in order to allow Congress time to act.

As a result of our review of the final EIS, we have reservations about the extent of the Commission's consideration of several environmental impacts of the proposed project. Our comments and questions follow.

1. Of primary concern are the effects of inundating over 40 miles of the upper New River, over 200 miles of tributary streams, and approximately 40,000 acres of field and forest land, eliminating a largely natural environment of high present recreational, aesthetic, and biological value.

Although EPA, in its comments on the draft EIS, specifically requested the Commission to discuss features of the New River that would make it desirable as a Wild and Scenic River, Section VI of the final EIS limits discussion of this important alternative to two sentences. The potential environmental and economic benefits of this alternative should have been presented in sufficient detail to allow comparison with, and subtraction from, the numerous benefits claimed for the proposed action. Has there been any subsequent investigation of this alternative?

2. The present productivity of the area's farms, forests, and streams should be better quantified and should appear as costs to the public where reduced or eliminated by the project. The costs and benefits of the project given in Appendix E were nearly five years old when the final EIS was prepared. Have these figures been updated and/or replaced by more current estimates?

3. The comments on the final EIS made by the EPA raise several questions about adverse effects resulting from project-induced development, both around the shoreline of the proposed reservoirs and in the flood plains below. These questions and the recommendations made by EPA concerning them deserve a considered response; has the Commission taken any action in this regard?

4. Our final concern is for the most efficient and conservative use of energy. Pumped-storage projects such as this one are designed to provide extra power during peak demand periods by using power generated elsewhere during slack periods; this process consumes approximately a third more energy than it produces. Has the Commission or its applicants considered alternative rate schedules to discourage these demands and encourage more uniform and efficient uses of power? The design and performance of primary generating plants could thereby be improved, and long distance transmission of power might be reduced.

We look forward to your response regarding these points and would be happy to meet with you on this matter to pursue the issues in more depth.

Reading this letter from the CEQ, I was extremely pleased. Its criticisms of the Blue Ridge Project were strong ones that could not be ignored by a court. I felt that with this letter and the new legal arguments we had developed, chances were good for a stay of the project if Congress did not act.

Yet all we had done, even if successful, would result only in a delay of the project. We wanted more than a delay; we wanted the New River legally protected against Blue Ridge and other water-resources projects. This could be done only through the river's designation as a scenic river. But if the scenic river study bill were defeated in Congress, delay would do us no good. We needed an alternative strategy to name the New a scenic river. Was there any way that would not require congressional action?

One day in early October, I had an idea that was so simple I wondered why no one had thought of it before. Reading over the National Wild and Scenic Rivers Act, I was reminded that there is an alternative method to establish a federal scenic river. A provision of the law, section 2(a)(ii), allows the governor of any state to apply to the Department of the Interior to include a river segment within the state as a federal scenic river. The governor must, in his application, certify (1) that the river segment has been designated a scenic river under state law and (2) that the state has adopted a management plan under which a state agency will manage the river without expense to the United States. Upon receiving such an application, the Secretary of the Interior is statutorily required to carry out a study to see if the river meets the criteria of a scenic river. If it does, Interior can legally establish the scenic river without any action by Congress.

This procedure seemed ideally suited as a wholly new strategy to protect the New River permanently against the dam. Some 4.5 miles of the part of the New that would be impounded had been designated a state scenic river in March of 1974. All that remained was to formulate a management plan and get the governor to make formal application to Interior. Such an application could accomplish all the scenic-river study bill pending before Congress was designed to do—authorize Interior to conduct a study on the New and thereby force the FPC to delay the January 2, 1975, effective date of the license.

There were some legal obstacles that had to be faced, however. For one thing, the 4.5-mile state scenic river was perhaps too small a segment to be eligible for designation as a national scenic river. Interior at that time had not adopted any formal rules on criteria for eligibility, but their informal guidelines stated that, to be eligible, stream segments had to be at least twenty-five miles long. This was qualified, however, by the statement that "a shorter river or segment that possesses outstanding qualifications can be included in the system." This obstacle, therefore, could possibly be surmounted.

Another problem was that the procedure outlined in section 2(a)(ii) had rarely been used before. It had *never* been used for a river, like the New, that had not first been designated by Congress under section 5 of the Wild and Scenic Rivers Act as a *potential* addition to the system. But, I thought, there is always a first time for everything. I could find nothing in the law itself that limited the section 2(a)(ii) procedure to rivers specifically named by Congress as eligible. It was worth a try.

Another obstacle to invoking the section 2(a)(ii) procedure was the fact that it was being done so late. The FPC license for Blue Ridge had been granted on June 14, 1974. Would Interior accept the application as valid if it was submitted just a few months before the license was to become effective? I did not know the answer but decided there was no choice but to go ahead. The matter as to whether the application was too late was one that Interior and the courts would have to decide.

I telephoned Ernie Carl and excitedly told him about the section 2(a)(ii) procedure. I said that I would prepare the letter of application for Governor Holshouser's signature and that the state should immediately start getting together a management plan for the New River. Carl agreed, and within four weeks his office put together a management plan and found several landowners who were willing to donate land or scenic easements in order to satisfy Interior's requirements for land protection.

Ironically, once we started preparing for a section 2(a)(ii) application to Interior, we discovered that, in March, 1974, Governor Holshouser had previously discussed this possibility in his conversations in Washington with Interior Secretary Rogers Morton and his aides. Morton, in a letter dated March 23, 1974, had even outlined the necessary steps. At that time, however, the governor's office had decided against trying to invoke section 2(a)(ii), believing that the state could not make a legally valid application and that it would be better to rely on passage of the scenic-river study bill by the Congress. By doing so, the state had passed up the opportunity of making application before the FPC license was granted. Even now, after the section 2(a)(ii) application had been prepared and was ready to send to Interior, the governor delayed signing it. He feared that it might not be acceptable and that it would interfere with the effort being made by Congress. Ernie Carl and I grew more and more distressed with each passing day. The later the application was made, the less likely it was that it would be viewed either by Interior or the courts as a bona fide application.

The situation finally came to a head on December 11, when the House Rules Committee voted thirteen to two against the scenic-river study bill. It was then apparent to us all that Congress would not act. This produced a crisis atmosphere in North Carolina. On December 12 Governor Holshouser sent the section 2(a)(ii) application to the Department of the Interior:

Mr. Rogers C. B. Morton
Secretary
Department of the Interior
Washington, D. C. 20240

Dear Rogers:
Pursuant to the authority of Section 2(a)(ii) of the Federal Wild and Scenic Rivers Act of 1968, 16 U.S.C. Section 1273(a)(ii), I hereby make application to you as Secretary of the Interior to designate the New River in north-western North Carolina as a "scenic river area" under the Federal Wild and Scenic Rivers Act.

In support of this application I hereby certify that the following actions have been taken by the State of North Carolina in order to comply with the conditions of Section 2(a)(ii):

1. The General Assembly of North Carolina has enacted legislation designating the Main Stem of the New River in North Carolina as a state scenic river area (G.S. 113 A-35.1) and has designated the South Fork of the New River for study as a state scenic river. (Resolution No. 170, Session Laws of N.C. 1973 General Assembly, 2nd Session 1974).

2. A management plan has been formulated by the Department of Natural and Economic Resources of the State of North Carolina for the permanent administration of the Main Stem and the South Fork of the New River in North Carolina as a scenic river. The implementation of

the plan will involve no expense to the United States. We are presently negotiating scenic easements and we have condemnation authority should this be required. The management plan is attached as Appendix A.

Pursuant to the Federal Wild and Scenic River Acts of 1968, you as Secretary of the Interior have the statutory authority to carry out a study to determine whether the New River in North Carolina meets the criteria set out in the Act for "scenic river" status. It is the position of the State of North Carolina, of course, that the New River possesses the required characteristics and we stand ready to cooperate with federal officials in aiding them to make the required study.

I'm aware that the Federal Power Commission has issued a license for a pumped storage–hydroelectric facility on the New River that would affect the branch of the New River which the General Assembly of North Carolina has designated a state scenic river prior to the issuance of the license.

I am today writing to the FPC to request them to further delay the effective date of the license pending your review of this request and the completion of the study which the Department of the Interior would carry out pursuant to Section 2(a)(ii) of the Federal Wild and Scenic Rivers Act. In the event the FPC refuses this request, the State of North Carolina will petition the Court of the District of Columbia to compel a delay of the effective date of the license pending the completion of the Section 2(a)(ii) study process.

Sincerely,
James E. Holshouser, Jr.

Although we were asking the FPC to delay voluntarily the January 2 effective date of the license, no one expected this to happen. Only a court order would stop Blue Ridge. In a meeting with state officials on December 11, I was formally asked to participate with the state attorney general to represent North Carolina in court, and we were authorized to take whatever court action was necessary in the matter.

The first priority was to prepare the motion for the Court of Appeals in Washington asking for a stay of the effective date of the license. We were in a race against time. It was already the middle of December, and we had to prepare the formal motion and a supporting brief. We would also have to ask the court for expedited consideration since, if the court handled the matter in the course of its ordinary work load, the case would not be heard before January 2. On Saturday, December 14, Millard Rich, an assistant attorney general, and I spent all day drafting the motion. That weekend I revised our brief, and on Monday, December 16, Ernie Carl hand-carried it from my office in Chapel Hill to Rich in Raleigh to be typed and filed together with the motion.

Then an absurd hang-up occurred. On Wednesday, December 18, I called Rich to see if the motion for a stay and the brief had been filed with the court. Rich replied that things in his office had been in a turmoil all week. He explained that arrangements had been made previously for painters to come in and completely redecorate the offices of the North Carolina Department of Justice. Everyone had expected a slow week just before Christmas. His secretary had not been able to start typing the papers, and he did not know when the painters would finish.

I hung up the phone with a queasy feeling. The next week was Christmas week. No one did any work then. The dark thought crossed my mind that the New River would go down the drain because the attorney general's offices were being redecorated! I picked up the phone and told Ernie Carl the situation. Carl personally found a typist and checked with her every hour on Wednesday and Thursday, riding herd until the papers were ready for filing. Finally, on Friday, December 20, 1974, the motion for stay was mailed to the court.

Asking the Court of Appeals in Washington for a stay of the license did not exhaust the possibilities we had of going to court. In the Court of Appeals case, North Carolina was contesting the decision of the FPC to grant the license for Blue Ridge. But the governor's section 2(a)(ii) application and the refusal of the FPC to extend further the effective date of the license while the Department of the Interior carried out a study of the New as a scenic river was a separate matter that we wanted to raise in court. Should this be argued before the Court of Appeals, or should we begin a new lawsuit to obtain a ruling on this point?

When I looked at the law on this question, I found no clear answer. The Court of Appeals has exclusive jurisdiction to review an FPC license, but judicial review of an agency decision under the Wild and Scenic Rivers Act is in the federal district court. Accordingly, I decided the best course of action would be to argue this point in *both* courts, disclosing to each that we were taking this action because we did not know which court had jurisdiction. Only in that way could we be sure to obtain a ruling on the matter.

I thus prepared a second lawsuit for filing in the federal district court in Greensboro, North Carolina, to obtain judicial review of the FPC's refusal to accede to the governor's request to delay the effective date of the license pending Interior's scenic-river study. The State of North Carolina would be the principal plaintiff, and it was decided to join as plaintiffs several of the landowners along the New River: Lester Halsey, Jim Todd, Jack Phipps, Lon Reeves, Hal Eaton, and Jese Colvard. The complaint, together with a motion to enjoin the Blue Ridge Project, was filed on December 19, 1974. We asked for a court hearing as soon as possible.

At the time the suit was filed, there was uncertainty over how the Department of the Interior would react. If Interior should reject Governor Holshouser's section 2(a)(ii) application, the state would have no alternative but to join Secretary Morton as a defendant. It was desirable to avoid this, since North Carolina needed Interior's support. Ernie Carl was negotiating for that support with Morton's aides. This resulted in a letter from Morton to Governor Holshouser on December 24, 1974, which gave unequivocal support to the state's position:

Dear Mr. Holshouser:

This responds to your December 12, 1974, letter requesting the designation of the New River in northwestern North Carolina as a component of the National Wild and Scenic Rivers System.

As you know, I have strongly supported the proposed legislation to include the New River in Section 5(a) of the Act for study as a possible component of the National Wild and Scenic Rivers System. Congress has not enacted the necessary legislation.

Now that you have requested Secretarial designation of the River it is necessary for us to make a quick study of the segment of the River identified for inclusion in the System. In addition, we must comply with Section 4(c) of the Wild and Scenic Act (P.L. 90–542) which requires that the Secretary of the Army, Secretary of Agriculture and the Chairman of the Federal Power Commission be given up to 90 days to review and comment on any applications to be acted upon by the Secretary of the Interior. We need at least six to nine months to study and make a decision on this matter.

Hopefully, circumstances will allow us this time.

> Sincerely yours,
> Rogers C. B. Morton
> Secretary of the Interior

This strengthened the state's hand, and the second suit also allowed North Carolina a separate chance to obtain a delay of the January 2 effective date in case the Court of Appeals did not act. Furthermore, it required the FPC and Appalachian to come into a North Carolina courtroom to defend their license.

Judge Eugene Gordon granted an oral hearing on North Carolina's motion for injunctive relief on December 30. I represented the plaintiffs, together with Millard Rich and Norman Smith, the senior partner of a Greensboro firm and one of the best attorneys in the state. The defendants brought a formidable array of legal talent. In addition to Steven Taube and other lawyers for the FPC, Appalachian intervened in the case, represented personally by the general counsel for American Electric Power Company, A. Joseph Dowd, and their Washington counsel, William Madden. In addition, they brought, as local counsel, two of the most respected members of the North Carolina bar, J. Ruffin Bailey and Kenneth Wooten, Jr. Appalachian and the FPC filed a motion to dismiss the case on the ground that exclusive jurisdiction lay in the Court of Appeals.

About an hour after the hearing began, an attorney for the Commonwealth of Virginia burst into the courtroom and demanded to be heard. He explained to Judge Gordon that he had missed his plane connection and was sorry to be late. He then presented a lengthy petition to intervene on behalf of Virginia. Having done so, he proceeded to argue that the action had to be dismissed because the litigation now involved a suit by one state, North Carolina, against another state, Virginia, and that only the Supreme Court of the United States had jurisdiction over such a case.

Judge Gordon was exasperated. North Carolina argued that jurisdiction resided in the district court, Appalachian and the FPC contended for exclusive jurisdiction in the Court of Appeals, and Virginia opted for the Supreme Court! Gamely, the judge asked to hear arguments on the merits of North Carolina's claim. The attorneys sparred back and forth on whether the Wild and Scenic Rivers Act required a halt to the project during the study period.

At the end of the argument, Judge Gordon announced that he was unable to come to a decision right away on either the jurisdiction or the scenic-river issue. He called for additional briefs from the parties and said he would take the matter under advisement.

Then the question came up whether the license would become effective as scheduled on January 2, 1975. Dowd tried to press Judge Gordon on this:

MR. DOWD: I'd like to add one thing to the January 2 date and the significance of it. On January 2, our license becomes effective. Now, this has nothing to do with construction, but it does mean that under the Federal Power Act, we have a valid, effective license pursuant to which we may proceed to carry out studies and expend the money subject to the possibility of the Commission's order issuing license being reversed on appeal to the Court of Appeals. In other words, we have on that day a contract with the United States Government.

THE COURT: All right.

MR. TAUBE: Your Honor, I'm hesitant as to what you have stated.

THE COURT: I have just stated that ultimately I gather that I will have to consider the matter with preliminary injunction. I will have to make a decision on that; and at that date, whenever it is, and hopefully it will be before long, the fact that the Defendants may have expended money or changed their situation in various ways in a manner that is hard to turn back, that will have no effect on me. For instance, in the New Hope Dam case, one other thing there that was brought to my attention—

well, if you allow them to go ahead and spend millions of dollars down here, to spend millions of dollars down here to build a dam and so forth, that no reasonable Judge is going to ask them to tear it down. I'm saying that if the dam is built when I get to a decision on this preliminary injunction, it is not going to make any difference to me. It will have to be torn down. You are proceeding at your own peril, in the meantime, until I enter some decree on this injunction.

Taube, on behalf of the FPC, made another attempt to defend the license:

MR. TAUBE: Well, I understand you, and I just feel it should be known on the record that the Federal Power Commission has been issued a license. The hearing was denied on August 12. There was a petition before the Court of Appeals. On January 2, that license becomes effective, and we in the protection of our license feel that the only place, the only one that it [sic] can ever affect the standards of that license is the Court of Appeals in the District of Columbia.

This caused an even stronger statement by the judge:

THE COURT: I'm not saying you're right or you're wrong. But they say you are violating this Title 16, Section 1278; and if I find they are right and that they are entitled to injunctive relief, I'm going to restrain it and enjoin you from proceeding, and that regardless of what you have done. If you have expended ten billion dollars—

Although this fell short of an injunction of the project, the North Carolina forces felt they had achieved a moral victory. The judge's attitude was a clear indication to the FPC and the power company that they could no longer count on the January 2 effective date. Their license would be in limbo for an indeterminate time.

Meanwhile, matters were coming to a head in the Court of Appeals. On December 31, 1974, Appalachian, fearing that a stay of its license was imminent, wrote to the Court of Appeals and voluntarily agreed to forego the assertion of any rights on the license until January 31, 1975. The Court of Appeals then denied North Carolina's motion for expedited consideration of the stay, since Appalachian's concession allowed it until the end of January to act.

Thus the new date for the license to take effect was January 31,

1975. We had to wait out the month, hoping that one of the two courts would act. Not only was court action necessary to prevent the license from becoming effective, but Interior took the position that it would not act on Governor Holshouser's section 2(a)(ii) scenic-river application unless Blue Ridge was enjoined by a court. On January 20, 1975, James G. Watt, director of the Bureau of Outdoor Recreation, the unit of Interior that would conduct the study, wrote to Edmund Adams:

The department now has under consideration an application from Governor Holshouser to designate a segment of the New River as a component of the National Wild and Scenic Rivers System pursuant to section 2(a)(ii) of the Wild and Scenic Rivers Act. We will be able to pursue our review of that application and to make a final determination, which involves consultation with other Federal agencies, only if the courts delay issuance of a Federal Power Commission license for the Blue Ridge Project.

Thus everything seemed to be on the line.

The decision went down to the last possible day. Finally, on January 31, 1975, the Court of Appeals handed down its decision. In a brief order, the court stated that North Carolina's petition for stay was granted and "the order of the Federal Power Commission under review herein is stayed pending the further order of this Court."

Another reprieve had been given to the New River.

8

The People Unite

THE LAST-MINUTE STAY OF THE BLUE RIDGE
Project on January 31, 1975, was obtained because all the resources
of the state government of North Carolina were mobilized to save
the New River. Governor Holshouser gave his personal attention to
the problem, and he said jokingly that, if all else failed, he would
call out the National Guard and mass them on the Virginia border
or carry out a sit-down against Appalachian's bulldozers. Everyone,
including Holshouser, laughed at the suggestion of the wag who
wrote that the state should simply divert the waters of the New
River before it entered Virginia, bringing them east over the Blue
Ridge Mountains and down into the North Carolina piedmont.
This would stop the dams, create a two-thousand-foot waterfall as a
tourist attraction, and keep all the water in the state. Short of that,
however, Holshouser was willing to try anything to keep the New
River running free.

Observers marveled that the New River issue had united two bitter political enemies, Holshouser and the state's attorney general,
Rufus Edmisten. A few months before, during the election campaign of 1974, there had been acrimony between the two men.
After Robert Morgan resigned as attorney general to run for the
United States Senate, Holshouser had appointed a Republican to
fill out the unexpired term. Edmisten had challenged the Republican appointee, and Holshouser had campaigned against Edmisten.
But the governor and the attorney general now cooperated fully in
taking action to protect the New River. Coincidentally, Edmisten,
like Holshouser, had personal ties to the town of Boone and liked
to reminisce about his childhood days spent playing along the river.

But if the political leaders of North Carolina were united, the people were not. The stay won in court had bought a little more time for the New River, but this would last only until the court got around to full consideration of the case. Nothing would be gained unless people worked hard against the dams during the time that had been won. But differences of opinion had developed among the people of the upper New River Valley. In 1974 an organization had been formed called the Ashe County Citizens Committee, whose purpose was to oppose the movement to declare the New a national scenic river. It adopted the slogan, "Dam the Scenic," which began to appear on car bumpers and roadside signs. This group also collected signatures of property owners along the South Fork who opposed the scenic-river designation. As a result, they claimed to speak for 74.6 per cent of the people, who together owned over fifteen thousand acres of river frontage. The group was strong enough to get Ed Adams, the Sparta attorney who was a longtime opponent of Blue Ridge, replaced as attorney for Alleghany County.

There were several factors responsible for the growing pro-dam movement in the valley. First and foremost, people were tiring of the long fight. The area had been under sentence of death for eight years. Three times during that period, the final decision had actually been at hand. It seemed pointless to oppose the inevitable. During the long period of indecision, many people had given up, sold their land, and moved out. Services were curtailed, as it was considered foolish to repair roads or improve communications in the area to be covered with water. Appalachian Power Company, through its special real-estate subsidiary Franklin Realty, had purchased many thousands of acres in the valley, and the old homes on them were simply left to rot in the wind and rain.

Second, most people did not have a clear idea what a "national scenic river" would mean for them or the valley. They thought of it as a large national park. This represented the intrusion of the federal government into the area. People were opposed to the power project, but they resented federal control of their lands even more. In addition, they had a natural distrust of politicians and wondered why so many were pushing the scenic-river bill. They also feared the outside environmentalist groups and thought that they would be restricted in the use of their own lands to accommodate hordes of canoeists and backpackers.

Appalachian did its best to play upon these fears and, in its local advertising, raised the specter of federal restrictions that would make farming impossible, remove lands from the counties' tax base, and turn the valley over to the Department of the Interior, the tool of the large conservation organizations. Once federal control was a reality, the scenario went, the big preservationist groups with their "paid full-time lobbyists . . . large budgets, and large memberships" would put the squeeze on the people of the valley and drive them off their lands in the name of recreation. In short, the scenic river would cost them their most cherished value, their freedom. People felt it would be better to sell out to the power company; at least they would be compensated for their lands.

The most outspoken local leader of the pro-dam forces was Robert Troutman of West Jefferson, in Ashe County, a wood-products manufacturing official and the owner of an eight-acre tree farm on the river. He called "all this historical crap" about the New River "ridiculous" and took the position that the dam represented the only chance the county would ever have for economic development. Another local businessman, J. C. Jenkins, a West Jefferson tire dealer, argued that the scenic river would do nothing more than bring in the state and federal governments. He called the opponents of Blue Ridge "just a bunch of meddling fools."

On the other hand, many people were still holding out against the dam. They realized that the only alternative to Blue Ridge was the scenic river. In some cases, family members were split in their attitude toward the scenic river. One family patriarch, the owner of a large dairy farm in Alleghany County, is said to have threatened his son with disinheritance unless he stopped his pro-dam activity. Others attempted to counteract Appalachian's advertising campaign. Everett E. Newman, Jr., expressed the sentiment of many people in a letter to Claude Kirkland, a division manager for Appalachian:

If you favor the Blue Ridge Project, as I know you do, then say so and please stop telling me, and others of the area, how good it will be for us. We have been doing fine for the past two centuries without having our land covered with water and I'm sure we will continue just fine in our present life-style.

It seems that you have put a price tag on everything. How you, or anyone else, can price a river is beyond me. I'm sure that the river was not meant to be priced when it was created.

I maintain that it is your company that has been putting out false information to sway public support to your views. I, for one, will not be swayed and will continue to do everything within my power to see that New River is preserved in its natural state. Also, I will continue to encourage others to do the same.

May New River flow free long after we both have departed this world.

Most of the people who still opposed the dam were the farmers of the valley who, throughout the period of uncertainty, had refused to give up or sell out. Twenty-seven-year-old Steve Douglas of Crumpler, in Ashe County, who farms the bottomlands of the South Fork that have been in his family for two hundred years, said: "What we've got here, money couldn't buy it. There's meaning to it. There's meaning to more than the dollar."

Up in Grayson County, Virginia, the foremost opponent of the power project was seventy-nine-year-old Cam Fields of Mouth of Wilson. Dressed like an Edwardian banker, he runs a general store, a Ford dealership, and a fifteen-employee woolen mill and has lived in the same house since 1907—all of which, as he pointed out, would be under 160 feet of water if Blue Ridge was built. He had fought Blue Ridge for many years without much hope of success but was buoyed by North Carolina's vigorous opposition. "We have no alternative except to fight," he declared.

Other leaders among the Virginia opponents were the Reverend Hal Eaton, a Baptist minister and head of Oak Hill Academy, a Baptist boarding school, and Glenn Halsey, a farmer and former member of the Grayson County Board of Supervisors. Eaton undertook to coordinate letter-writing and the gathering of names for petitions against Blue Ridge. Halsey vowed to fight to the finish: "We're going to give 'em everything we got."

The opponents of the Blue Ridge Project coined the term "beaver" to refer to the pro-dam people, and this became the name applied derisively to anyone who had anything to do with the Ashe County Citizens Committee. They also looked for a counterslogan to "Dam the Scenic." Recognizing that the people of the New River Valley just wanted to be left alone, the Blue Ridge opponents decided on the rallying cry, "The New River like it is."

It would take more than a new slogan to stop the Blue Ridge Project, however. The people in the upper New River Valley would have to join forces with the North Carolina state government to form a united political movement against the dams. They recognized that a new organization had to be formed for this purpose.

The first effort to create a new, statewide organization to fight for the preservation of the New River was made in October, 1974. Joe Matthews and R. Edwin Shelton of the Northwest Environmental Preservation Committee, based in Winston-Salem, contacted John Curry, the president of the Conservation Council of North Carolina, for the purpose of arranging a rally to muster support for a new coalition to preserve the New River. They arranged a meeting at Elk Shoals Methodist Campground in Ashe County, on October 20, 1974. A program was scheduled featuring a slide show and speeches by Ed Adams and Lorne Campbell. Invitations were sent out to members of both organizations and other interested parties.

This new effort got off to an inauspicious start. The owner of the campground became intimidated at the last minute and posted a No Admittance sign. The local West Jefferson radio station broadcast messages of alarm that all citizens of the area should beware of motorcycle gangs and contingents of Hell's Angels from outside the county who were allegedly headed for the campground to take part in a rock orgy to save the New River! This successfully foiled the attempt to begin a new grass-roots campaign.

But the organizers did not give up. In December, 1974, Curry, Matthews, and Adams decided to make a new attempt to create a specific organization for the single purpose of raising money and carrying on a political-action campaign to save the New River. Another meeting was called, for January 4, 1975, at Poe Hall on the campus of North Carolina State University in Raleigh. Invitations were sent to members of North Carolina–based conservation organizations: the Conservation Council of North Carolina, the Upper New River Valley Association, the New River Chapter of the Izaak Walton League, and the Northwest Environmental Preservation Council.

Citing the court actions begun by the State of North Carolina, the invitation stated that "there is still time to save our ancient and priceless treasure, the New River." The objectives of the new organization were to be (1) support for additional North Carolina legislation to extend the portion of the New River in the state scenic-rivers system to include the entire South Fork, in addition to the already included 4.5 miles of the main stem, (2) support for the passage of scenic-river legislation in the U.S. Congress, and (3) education as to the plight of the New River. A card was enclosed with each invitation so that people unable to come to the meeting could indicate their support and get on the mailing list of the new organization.

This meeting was a great success. Interest was running high, since the adverse vote in Congress and the court actions had been given broad publicity by the news media of the state. Over two hundred fifty people attended, and the new organization, christened the Committee for the New River, was formally begun. Ham Horton, the former North Carolina General Assembly member who had pushed the early state scenic-river legislation, was elected president. Joe Matthews was elected executive secretary. Ed Adams and Lorne Campbell were chosen vice-presidents, Polly Jones became secretary, and Louise Chatfield, treasurer. John Curry agreed to serve as legal counsel.

This was a formidable group of people. Ham Horton was a widely respected political figure with an intimate knowledge of state politics and strong ties to the North Carolina congressional delegation. Joe Matthews was an experienced administrator with the ability and willingness to carry on the day-to-day operations of the committee. Ed Adams and Lorne Campbell were two local people of the valley who had carried the opposition to Blue Ridge on their broad shoulders for a long time. Polly Jones, a housewife from Ashe County, turned out to be invaluable in educating and convincing her friends and neighbors, who seemed to be practically everybody in all three counties. Louise Chatfield of Greensboro had long been an active conservationist. John Curry was a natural leader and an able lawyer.

The Committee for the New provided a focus for a tremendous outpouring of support. Lorne Campbell went back to Grayson County and coordinated the Virginia members' activities. Within a few days, the fledgling organization had seven hundred members in five states and was growing daily. The amalgamation of local persons and outside supporters created a powerful political force. Through Ernie Carl, the activities of the new group were coordinated with the efforts of the State of North Carolina.

Meanwhile, Carl, Art Cooper, and Secretary Harrington got the state government moving on the study the North Carolina General Assembly had authorized in 1974 to extend the 4.5-mile scenic-river segment of the New by adding the South Fork to the state scenic-rivers system. In December, 1974, Robert Buckner of the Office of Recreation Resources conducted a study of the South Fork, collected technical data, and carried out a field examination of this section of the river. After assessing its physical characteristics and the recreational alternatives, he concluded that it met the criteria for scenic-river status.

On the basis of Buckner's report, Carl set up a series of public hearings to be held on the new scenic-river proposal, in Sparta and Jefferson on January 27, 1975, and in Boone on January 28. These were used to publicize the scenic river and to allay the fears of the local people as to what would happen to their lands. Publicity for the public hearings was coordinated with the Committee for the New River. In its first newsletter, the committee included a fact sheet and asked for support. It also distributed three thousand bumper stickers—"The New River Like It Is"—and made a slide program on the scenic river available for the asking. The newsletter asked North Carolina members to contact their own representatives in the General Assembly to get their support for the bill to extend the state scenic river. It also asked for donations, stating, "We're almost broke. Don't worry though: we won't let it slow us down."

Art Cooper personally presided over the two days of hearings. It was a wild scene, packed with both supporters and opponents of Blue Ridge. Cooper simply and calmly explained the concept of a scenic river, saying that there would be minimal restrictions on traditional farming and forestry activities and that only a few hundred acres would be acquired by the state. The primary management tool would be the purchase of scenic easements along the river, which would still permit private use of the lands. These hearings, the first that any governmental agency had held relating to the New River in the area actually concerned, did much to allay suspicion of the scenic-river concept and began to convince people that a scenic river was preferable to the power project.

It was necessary for North Carolina to extend state scenic-river status to include the South Fork because officials of the Department of the Interior felt uneasy about processing the governor's section 2(a)(ii) application of December 12, which concerned only 4.5 miles of the main stem of the New River. Interior told North Carolina that it supported the study application but that it would be best if the state could pass legislation to extend the number of miles of river to be included. On January 15 and 16, 1975, North Carolina Senator Jesse Helms and Congressman Steve Neal, a Democrat who had defeated Mizell in the 1974 election and who now represented the district that included Ashe and Alleghany counties, introduced scenic-river study bills in the new session of the U.S. Congress; but the focus of activity in early 1975 was to gain the necessary study by Interior through the application by Governor Holshouser. With the stay by the Court of Appeals on January 31, 1975, there was time to shepherd the needed legislation through the North Carolina General Assembly. The strategy developed was first to concentrate on the South Fork scenic-river bill and then to

again ask Interior to protect the river by declaring the New a federal scenic river.

The January hearings were followed by two months of hard lobbying of North Carolina legislators by the Committee for the New
and state officials. The lawmakers had to be convinced that they
would gain politically by supporting the bill. The Ashe County Citizens Committee was active in opposition. The major North Carolina
electric utility companies, Duke Power Company and Carolina
Power and Light, instructed their lobbyists to stay neutral; they had
their own battles to fight and did not want to cross swords over
such an emotional issue. The Committee for the New had people
throughout the state ask their own individual legislators for commitments of support. Organization paid off, for with its statewide
membership base, the Committee for the New was able to surpass
by far the effort made by the essentially local Ashe County Citizens
Committee.

Appalachian was running into trouble elsewhere too. In February
the West Virginia Public Service Commission denied the power
company all but 1.5 per cent of a 10.5 per cent rate increase it had
put into effect in 1971. Appalachian was ordered to refund $39.3
million to its West Virginia customers. Power-company officials
announced that the utility's financial condition might require it to
defer some of its investment plans. In addition, North Carolina
State Senator McNeill Smith of Guilford County introduced legislation to require the North Carolina Utilities Commission to study
peak-load pricing for electric utilities. This would not have affected
Appalachian directly, since the company did not sell power in North
Carolina. It did, however, help convince North Carolinians of
the validity of one of the chief arguments against Blue Ridge: that
peak-load pricing was feasible and could make the project super-
fluous. On June 18, 1975, the Virginia Corporation Commission
announced its determination to review all electric-rate structures
to determine the feasibility of peak-load pricing. This would potentially affect Appalachian and Blue Ridge.

By April, 1975, the North Carolina Department of Natural and

Economic Resources had formulated a concrete plan for extending the state scenic-river status of the New. It had decided to recommend the South Fork from Dog Creek to Twin Rivers, which, together with the 4.5 miles of the main stem, would total 26.5 miles, barely enough to qualify under Interior's requirement of at least 25 miles. To gain support from local landowners, the plan called for outright acquisition by the state of two to four hundred acres for four recreational areas. No bridges would be removed and no footpaths would be constructed along the river. Farming activities were not to be disrupted; only constructing buildings or developing land within the area of scenic easements along the shore would be prohibited. Management of the area would be carried on, not by the federal government, but by the state. To nail down local support, Art Cooper held another public meeting, on April 8, at Ashe Central High School, to explain the plan to local landowners.

By the efforts of the Committee for the New and state officials, local support for the scenic river grew, and the state's legislators were lined up to vote for the necessary legislation. Valuable outside help came as well. Earl Hamner, Jr., a native Virginian and the creator of the popular television series, "The Waltons," wrote a short piece that he entitled "Save the New River":

I am a Virginian, presently residing in California. It has been my good fortune during my life to know many of the world's great rivers. Last spring I sailed down the Nile from Luxor, Egypt to the Valley of the Kings where the great pharaohs were entombed. I have waded in the Mississippi at Davenport, Iowa and sailed in a pilot boat along its delta country above New Orleans. When I was a soldier in World War Two I was stationed in France and I sometimes went fishing in the Seine and some nights strolled along the Seine with a girl named Madeleine. There is a river called the Rockfish in my beautiful Nelson County which is known all over the United States and in some thirty-four foreign countries because I have mentioned it frequently on a television series called "The Waltons."

A river is a holy thing, and to walk to the side of a river bed, to sit and contemplate the flow and life and meaning of a river, no matter what size, has always been an almost mystical experience to me.

Looking out of my window I can see a river, but no one would know it for what it is. This river has been tamed, constricted to a concrete bed, emasculated and robbed of vitality. In the summer months it trickles along with the flow of a dying creek in its concrete banks. Even then it does not slow for the birth of tadpoles or minnows nor invite a flock of ducks to settle there. It is an antiseptic river, free of germs, free of vitality, free of any life.

Once in awhile, during the great desert storms of winter, the Los Angeles river regains some vitality. It becomes a surging giant and carries away all the debris (mattresses, sofas, worn out refrigerators, beer cans and garden clippings) which have been thrown into its concrete bed during the past year. But it can never again be a real river. It can never be free. I point it out to out-of-town visitors sometimes, and they laugh and say "You're not serious!"

Sometimes when I am homesick for Virginia, I look at the Los Angeles River and wish that it might be lined with willows and that a green bank where some wild honeysuckle lives could be there. I wish that I might walk along its edges and find some shady place where I could stick a fishing pole in the mud and sit there and wait for a catfish to come along. But there are no catfish, no bull-frogs, no cray-fish, carp, bass or perch, no fish at all, and if I were to be foolish enough to sit beside the Los Angeles River's concrete banks, I rightfully would be laughed out of town.

What people forget is that rivers are like dinosaurs, destroy their habitat and they become extinct.

We need rivers, not only for their commercial value, but for their beauty and their mystery and even for their occasional destructiveness to remind us that they were created by a power greater than ourselves.

I plead with those of you who have the power and need to "modify" the New River to come to Los Angeles and share the monotony of the Los Angeles River. Once it was alive. Now it is a sluggish stream contained in concrete walls of no use to anyone nor can it ever be of use again. Men have tampered with nature and in so doing killed a living natural wonder, but this happened in California. I hope the people of Virginia and North Carolina will have better sense.

Please think twice before you destroy the New River, a living thing that belongs to each of us. If you choke it with dams how will your children ever know what it once was, and how can you face them?

At the request of the Committee for the New, Hamner made a tape of this message so that it could be broadcast on radio stations throughout the state. It was very effective in gaining support for the scenic-river bill.

On April 22, 1975, identical bills were introduced in the North Carolina Senate and House "to lengthen the segment of the South Fork, New River, in Ashe and Alleghany Counties included in the North Carolina Natural and Scenic River System."

Third-grade schoolchildren at Summit School in Winston-Salem helped convince six members of the House and four in the Senate who were still uncommitted. Jennifer Cleland wrote to one: "Our river is something you should be proud of, I should think you'd like it much more than a noisy, dirty, ugly, horrible, etc, etc, etc, electric plant. If you are like most people, you will vote for the river's freedom. If you don't you'll be as depressed over it as I'll be mad over you. Help us, we're in this nation too." Emily and Susan wrote: "Rivers can't be replaced! But the hidroelectric company can be put in another place! We don't want to look with our face if theirs a company in that place! Rivers are fun to play in the sun!" And Molly Bass put it: "The whole Summit School I meen the whole Summit School wants the river saved!"

The schoolchildren got their wish. In May both houses of the North Carolina General Assembly passed the bill by unanimous vote, and on May 26 it became effective as a ratified bill. Now that the North Carolina General Assembly had been won over, it was time to turn to Interior.

An unforeseen problem had arisen, however. The Secretary of the Interior, Rogers Morton, who through Holshouser's efforts had supported federal scenic-river status for the New, resigned in April, 1975, to accept a new job as Secretary of Commerce. This created uncertainty over what position Interior would take on the New River in the future. To succeed Morton, President Ford nominated Stanley K. Hathaway, a former governor of Wyoming, who was generally regarded as a foe of environmentalists.

Hathaway's nomination to the sensitive post at Interior caused great dismay among the national conservation organizations. They regarded his record while governor of Wyoming as disastrous. The

Sierra Club, Audubon Society, and other well-organized groups decided to fight tooth and nail and to do everything they could to block his confirmation in the Senate. During his confirmation hearings before the Senate Interior Committee in April and May, Hathaway was the subject of extreme vilification.

It was Ernie Carl's job to analyze the changing situation at Interior to determine the impact on the New River. He determined that Hathaway had the votes and that though there might be a long fight, he would be confirmed eventually. Accordingly, Carl advised Holshouser that the effort to convince Hathaway should start right away, without waiting for Senate confirmation. This strategy called for Holshouser to ingratiate himself with Hathaway by becoming a strong supporter of his nomination. Holshouser met with Hathaaway, got in touch with the Interior Committee, and testified in favor of confirmation. This made an impression on the Committee and made Hathaway extremely grateful to Holshouser.

In June, just after Hathaway had been confirmed and took office, Governors Godwin of Virginia and Moore of West Virginia began to lobby Hathaway on behalf of Appalachian to turn down North Carolina's scenic-river application for the New River. They met with him personally and wrote him that "the segment of the New River which is the subject of the request does not meet the legal requirements of the wild and scenic rivers act." At that point, however, Hathaway's feelings toward Godwin and Moore were in the nature of "Where were you when . . . ?"

In June, 1975, North Carolina prepared a revised management plan for the enlarged state scenic river and a new application to be submitted to Interior asking federal scenic-river status for the 26.5-mile stretch of the stream. It was decided that, in formal terms, the application should be submitted to Interior as an amendment of Governor Holshouser's application of December 12, 1974. Since the timing of the request might become an issue, casting it in the form of an amendment to the original application would preserve the claim that a valid application had been submitted before the effective date of the license.

When the amended application was ready, it was decided that Holshouser should personally deliver it to Hathaway and ask his acceptance of it. On July 15, 1975, Holshouser obtained assurances from Hathaway that he would accept the request and begin the formal statutory review process required before final designation could take place. Holshouser and other North Carolina officials were elated. Victory was at last in sight.

Then another strange twist of fate intervened. Hathaway had been exhausted by the long-drawn-out confirmation process. He was mentally depressed. On the day after his conference with Holshouser and his decision on the New River, he entered a hospital, suffering from exhaustion and mental strain. In North Carolina, it was wondered whether conflicting pressures on him about the New River had been the "straw that broke the camel's back." On July 25, 1975, Hathaway submitted his resignation as Secretary of the Interior.

With Hathaway gone, there was again uncertainty over whether Interior would act on North Carolina's scenic-river application. Policy would be allowed to drift until a new secretary was confirmed. It was extremely discouraging for the North Carolinians. They had worked hard and covered every possible contingency except another vacancy at Interior.

Still, there was cause for optimism. Blue Ridge was stayed in the courts, North Carolina had been able to take the necessary action to strengthen its scenic-river application to Interior, and people across the state as well as in the affected counties were united as never before in their opposition to Blue Ridge.

The Committee for the New River had grown to a three-state organization. It was still North Carolina–based, but strong chapters had been organized in Virginia by Hal Eaton and in West Virginia by Jim Watkins of Beckley. After the successful passage of the enlarged scenic-river bill by the North Carolina General Assembly, the Committee for the New decided that what was needed was a large rally to emphasize the virtues of the river and to serve as a focal point for the gathering support. They called the celebration the Festival of the New River and scheduled it for July 26.

Polly Jones of Crumpler was appointed head of the organizing committee for the festival. She enlisted hundreds of people in Ashe County to share in the preparations. Everybody was given a job to do. Virtually every freezer in Ashe County was commandeered to store ice. Ham biscuits, hot dogs, corn on the cob, and cakes were prepared for the occasion. Shatley Springs sent over a truckload of spring water. Lon Reeves allowed his hayfield by the river to be the site of the festival, and farmers all along the river helped to harvest and bale the hay to prepare the field for use.

The necessity for this advance planning and cooperation did as much as anything to unite the local people against the dam. Those who had remained uncommitted became caught up in the preparations for the festival through Polly Jones. Once they had agreed to help and found that everyone else was doing so too, they no longer hesitated to express their opposition to Blue Ridge. The local "beavers" were now an isolated, quiet minority.

The Festival of the New was a huge success. It became a symbol, an expression, and a celebration of the river. Over three thousand people attended and listened to music, poetry, and a specially prepared historical drama about the people of the river. In addition to bluegrass and country music, six original songs were performed about the meaning and beauty of the river.

Elizabeth McCommon, a folk singer and actress who had gone back to the land to live on a virtually self-sufficient farm in Floyd County, Virginia, performed her "Ballad of the New River," a song of striking beauty:

> How rich, how green was my valley;
> that precious place my New River home.
> I could hear the torrent roar,
> I could see the high hawk soar
> from that spot of land my father gave to me.
> But it's gone forever more,
> sunk beneath that muddy shore,
> and a dream is dead for children yet unborn.

The Ballad of the New River
"That's What They've Said So Many Times Before"

© Elizabeth McCommon 1975
Floyd County, Virginia
U.S.A.

How rich, how green was my val-ley; that pre-cious place my New Riv-er home.__ I could hear the tor-rent roar, I could see the high hawk soar from that spot of land my fath-er gave to me.__ But it's gone for-ev-er more, sunk be-neath that mud-dy shore, and a dream is dead for chil-dren yet un-born.__ I must not cry, you see, pro-gress is good for me — at least that's what they've said so man-y times be-fore.__

continued —

Broad and brown moves the river; silent in sorrow it flows
No more sparkle flash or fun; to the dams it just gave in
They broke its back and mine is soon to go. Gone forever more
Sunk beneath that muddy shore, and they say the river west of here is dead
We need the power too, that the dams will bring to you. That's what
They've said so many times before.

North flows the Nile and the New; their ages unreckoned, unknown.
I guess they're old as Mother Earth; for age on age they've spanned her girth
Now dams sap the lives of the two. They are gone forever more.
Sunk beneath that muddy shore, say "hello say goodbye", sing "so long".
They brought more jobs more dough - I'll conceed it & then I'll go because That's what
They've said so many times before. (repeat first verse)

The Festival of the New, July, 1975. Crosspiece shows the depth of the proposed lake at this point. (*Courtesy of the* Winston-Salem Journal & Sentinel)

I must not cry, you see, progress is good for me—
at least that's what they've said
so many times before.

Ronnie Taylor of Fleetwood, North Carolina, performed another
song of protest, called "Mouth of Wilson Town," about the power
company's plan to drown the tiny Virginia town of Mouth of Wilson under 160 feet of water:

I ran into a friend of mine
A walkin down the road the other day.
He said have you seen the valley full of water
Down there where I used to play?
Still in my mind I see it now
The sky so blue the sun so far away.
But they've torn the old homeplace down
It's goin under fifty feet a day.

When I was a little boy I wandered
In green fields lost in time,
The sky was blue the sunshine through
Most all of the time,
But its a hundred feet and rising
In the memory of my mind.

When I grow old I am told I'll understand
Those questions in my mind.
I'll sit down in an easy chair those
Questions will be answered there in rhyme.
But it's two hundred feet and rising
In the memory of my mind.

Yes the asphalt on the highways
And the shade upon the skyways
Brings me down.
And it's two hundred feet and rising all down around
Mouth of Wilson Town.

A diverse group of people attended the festival. There were numerous farm families from the valley and from all over Watauga,
Ashe, and Alleghany counties in North Carolina and Grayson Coun-

ty, Virginia. There were environmentalists from statewide and national conservation organizations, political figures such as Congressman Stephen Neal and Ken Hechler, and craftsmen such as Dave Sturgill of Piney Creek, who builds guitars and dulcimers. All were brought together to make common cause for the river. As Ham Horton put it, "This is a people's movement. The beauty of it is it's purely homegrown. The people who are involved are mostly people who haven't been traditionally political activists. Most of the people here today plus the musicians and craftsmen are local people. They're here to preserve their roots."

The Festival of the New was an important step in the fight to save the river. It brought people together as never before. It was a celebration of the values the river represented, the stable, traditional way of life in the valley. The time had not yet come for a celebration of victory, however. Difficulties still lay ahead.

9

False Hopes and a New Defeat

IN THE LATE SUMMER OF 1975, AFTER THE FESTI-
val of the New, we all believed that the decisive phase of the struggle
to preserve the New River had at last begun. We felt that there was
reason to be cautiously optimistic. If the Department of the Interior
declared the New a federal scenic river, the power project would
have to be dropped.

But it was not quite that simple. The Federal Wild and Scenic
Rivers Act specified that Interior must, before approving the New
as a scenic river, submit the proposal for comment during a ninety-
day period to the Secretary of the Army, the Secretary of Agriculture, and the chairman of the Federal Power Commission. All three
agencies could be expected to oppose scenic-river status for the New
River. This meant there would be no quick decision on the governor's application, and the opponents of the scenic-river designation would have the time to muster their forces and make another
attempt to stop it. The more immediate problem, however, was to
get Interior's agreement to accept the governor's application and to
begin the study of the New River.

After Hathaway's resignation, the acting Secretary of the Interior was D. Kent Frizzell, a lawyer who had previously served as
solicitor for the agency. Governor Holshouser's office pressured him
to act, and the Committee for the New called on its members to
write letters asking Frizzell to proceed immediately with consideration of North Carolina's scenic-river application. Inside Interior,
key officials allied with North Carolina were also pressuring Frizzell.
On August 7, 1975, Deputy Assistant Secretary Douglas Wheeler
met with Frizzell and told him there were no political or legal reasons for waiting to start the review process. He argued that policy

had already been set under Secretaries Morton and Hathaway and that Frizzell, by delaying, would affront the State of North Carolina.

Frizzell, however, refused to be staked out on the issue. He was in a tenuous position. He wanted to be nominated to the position of secretary and in fact seemed to have the inside track on being named by President Ford. He hesitated to make any decision on a controversial matter that might turn important political forces against him. He was also sincerely uncertain as to how to proceed on the matter.

Accordingly, he delayed taking action one way or the other. Legal opinions were requested from Interior lawyers as to whether the department should wait until the courts had acted before doing anything to process the application. In addition, a group of advisers told Frizzell that the department should prepare an environmental assessment and an environmental impact statement before beginning the ninety-day review period. Governor Holshouser made a trip to Washington to try personal diplomacy to get Frizzell to act. This time he was unsuccessful, and Frizzell announced that he would take no action until after receiving the opinions of the legal staff on possible alternatives. During this period, Interior was deluged with several hundred letters urging Frizzell to act.

Meanwhile, the State of North Carolina was also playing for the time Interior needed. There were developments on the legal front. On May 7, 1975, Judge Gordon rendered a decision in the lawsuit the state had begun in federal district court. He refused to decide whether Governor Holshouser's scenic-river application to Interior was legally valid. He took the position that his court had no jurisdiction in the matter and the validity of the application was up to the Court of Appeals in Washington.

We decided to appeal this decision to the Court of Appeals in Richmond because we were not sure that the Washington court would agree to take jurisdiction despite Judge Gordon's opinion. We felt that it was essential that one of the courts directly involved decide the legal question of the validity of the scenic-river application.

In early August the state received a briefing schedule and a fall,

1975, date from the Court of Appeals in Richmond for the appeal from Judge Gordon's decision. The Court of Appeals case in Washington was also scheduled for argument in the fall. This meant that both court cases would be decided virtually simultaneously and would be over in a few months, probably before Interior had a chance to act! This would present no problem if either court ruled in favor of North Carolina's position. Then the power project would be remanded to the FPC. But if the state lost in both courts, the result would be potentially disastrous. Interior would probably abandon its efforts, and the state would run out of courts and legal arguments to block Blue Ridge. The license would take effect and there would be nothing further that North Carolina could do.

Thus Smith, Rich, and I decided to file a motion in the Richmond case to ask for a stay of the proceedings in that court until such time as the Court of Appeals in Washington had rendered its decision. This meant that even if the Washington court were to rule adversely, the state would have another legal forum to turn to in the case against Blue Ridge, giving Interior more time. The legal grounds for a stay were persuasive. The appeal in the Washington court involved judicial review of the license granted to Blue Ridge by the FPC. It also involved a challenge to the adequacy of the FPC's environmental impact statement on the ground that the statement did not consider the peak-load pricing or scenic-river alternatives or adequately disclose the costs and benefits of the power project. In the Richmond case, however, the state was asking that the court uphold the validity of the scenic-river application under the Wild and Scenic Rivers Act and tell the FPC that it acted illegally by refusing to delay the effective date of the license until Interior had a chance to complete its ongoing consideration of possibly designating the New a national scenic river. The FPC and Appalachian had, however, challenged the jurisdiction of Judge Gordon's court and the Richmond court to decide the issue of the delay of the effective date under the Wild and Scenic Rivers Act. The state, in its briefs filed in the Washington court, raised *this same issue*, stating its position that jurisdiction was more properly in Judge Gordon's court than in the Washington court; it did ask that a determination on the matter be made by one of the two appellate courts.

In these circumstances, North Carolina suggested, it would be in the interest of judicial economy for the Richmond court to wait for the decision of the Washington court. If the latter court were to reverse the FPC's grant of the license, the case in the Richmond court would become moot. On the other hand, if the Washington court were to consider the question of jurisdiction over the Wild and Scenic Rivers Act claim, the possibility of conflicting decisions in the two courts would be raised if the Richmond court was simultaneously proceeding to judgment on the same issue. It was even conceivable that each court could conclude that the other court had jurisdiction! Or, both courts could acknowledge jurisdiction over the scenic-river claim and proceed to conflicting decisions on that question.

On August 26, 1975, the Richmond court agreed with North Carolina's position and ordered the case stayed until a decision had been reached by the Court of Appeals in Washington. This was an important victory for the state and assured that Interior would have a little more time to act. It also helped to persuade Interior officials that they should not delay processing North Carolina's application while waiting for the outcome of the Washington court appeal.

Another development also helped North Carolina. On September 9, 1975, President Ford named Thomas S. Kleppe, a former Republican congressman from North Dakota and the head of the Small Business Administration, to the post of Secretary of the Interior. This was a disappointment to Frizzell, but it freed him from the political pressures swirling around him. He decided to act on the New River question. On September 12, 1975, he wrote Governor Holshouser that Interior would support the scenic-river application. He then ordered the Bureau of Outdoor Recreation within the department to begin studying the state's application and to prepare an environmental impact statement on the scenic-river preservation plan. Once the review had been completed and the impact statement prepared, the ninety-day review period could begin. Frizzell properly cautioned, however, that no final decision could be made until the study had been completed and the comments of other agencies evaluated.

Frizzell also asked North Carolina officials to clarify or update certain points involving the state's management plan for the river. Noting that the plan stated that $1 million would be requested for land acquisition and $5 million for public recreation sites state-wide each year of the 1975–77 biennium, he asked that the state specify how much of this would be allocated to the New River. Frizzell also asked about the status of floodplain zoning in Ashe and Alleghany counties, since the management plan called for zoning to regulate building activity on the banks of the river.

Action was thus held up for another month until, on October 13, 1974, North Carolina gave Interior acceptable assurances about these points and the state's ability to implement the management plan. By this time, Thomas Kleppe had been confirmed as Secretary of the Interior. In contrast to Hathaway's ordeal, Kleppe's confirmation hearings in the Senate were rather mild. Since he had never been involved in natural-resources issues, he had no track record that his enemies could use against him. Furthermore, the Senate Interior Committee, with Hathaway's breakdown on their minds, were in no mood to put Kleppe through the wringer.

While North Carolina's application was slowly inching its way through Interior, there was an important new development. In late August, 1975, Bob Poole, the Washington correspondent for the *Winston-Salem Journal*, was given a tip by Dr. William Gardner, chairman of the anthropology department at Catholic University in Washington, D.C., that Appalachian had had archaeological surveys made in the New River Valley and was withholding them from the FPC. Poole, together with Ned Kenworthy of the *New York Times*, investigated and found that Appalachian, at the FPC's request, had in fact commissioned two archaeological surveys of the valley in connection with the Blue Ridge licensing proceeding. One of these had been completed in 1965 by Dr. Harvard Ayers of Appalachian State College, and the other had been carried out in 1969—after Appalachian decided to double the size of the Blue

Ridge Project—by Dr. Charlton Holland of the University of Virginia.

Poole went to Charlottesville and interviewed Holland, who gave him a copy of his study. Holland believed, on the basis of his survey findings, that the New River Valley had been continuously occupied by man since at least 8000 B.C. and that the New River, as the only river bisecting the Appalachian Mountains, had been an important passageway for early man in North America. During this time the area had "received an unusually large number of cultural influences from many directions and was truly a cultural crossroads." Holland and other archaeologists feared that the Blue Ridge Project would inundate the valley before the area could be excavated: "If you go on the things that have been found there, the whole history of Indians in the eastern United States had its chapters in this area. It is absolutely important to know what is under that New River Valley." Holland explained that his two-week survey in 1969 located 42 archaeological sites, recovered 1,459 pieces of pottery and 415 arrowheads and other stone artifacts; he found one "very large Indian village" and several other "archaic encampments." The artifacts and other materials were all picked up off the ground without excavations.

Poole then checked with Appalachian to see what had happened to this information. He called Paul Johnson, the company's project engineer for Blue Ridge, who admitted, "I looked back in the files, and this was never transmitted to the FPC." He offered as an excuse the contention that Appalachian was trying to protect the sites from depredation and so decided to keep them secret. Poole thought this a pretty lame argument, since Appalachian was planning to flood the whole area as soon as possible.

Officials at the FPC could not recall seeing the surveys. They were not exhibited at any of the hearings on the license. When told about the surveys, Judge Levy, who had conducted the hearings, declared that the FPC "has no expertise in this area." The only provision that had been made by the FPC for archaeological investigation was a condition in the Blue Ridge license requiring Appalachian to spend up to $7,500 for survey and salvage before filling the reservoirs. Levy indicated to Poole that this was merely a standard, perfunctory provision. Holland called this sum totally inadequate, asserting that an adequate effort would cost about $40,000 and involve three or four years of work.

Poole published a newspaper article about his investigation in the *Winston-Salem Journal* on August 31. This alerted the attorneys for the State of North Carolina, who decided to bring the matter to the attention of the Court of Appeals in Washington. Rufus Edmisten, attorney general of North Carolina, vowed to use the newly discovered Indian arrowheads as "weapons" against the FPC.

Millard Rich and Ed Adams came to my office on September 1 to plan the legal strategy. The case in the Court of Appeals in Washington had been completely briefed by both sides and was set for oral argument on October 23. The evidence as to the archaeological surveys had never been raised, either before the FPC prior to its issuance of the license or before the court. We wanted to call the court's attention to the newly discovered factual evidence and ask the court to remand the case to the FPC so that the evidence could be presented to the agency.

Accordingly, we decided to utilize a procedure outlined in the Federal Power Act to get the court to consider the issue. A provision of that act says that a party may ask the Court of Appeals for "leave to adduce additional evidence" before the FPC if it shows that (1) the evidence is material and (2) there were reasonable grounds for failure to present the evidence in the original license proceeding before the FPC. In order to comply with this provision, Rich was dispatched to get affidavits from Poole and Dr. Holland. On September 11, 1975, North Carolina filed an Application for Leave to Adduce Additional Evidence, together with affidavits of Poole and Holland, with the Court of Appeals. Copies of the two archaeological surveys were made available to the court, and the application stated:

Both the Ayers and Holland Surveys were preliminary archeological field work to determine whether sites relating to the activities of early man were present within the Blue Ridge Project area. Both surveys show the existence of several dozen such sites and make recommendations for their excavation. Furthermore, both surveys show from a preliminary analysis of the artifacts, projectile points and pottery sherds found at the sites, that the New River Valley is one of the most important archeological areas in the eastern United States (Holland Affidavit). The New River is one of the oldest rivers in the world; it existed in pre-historic times as the Teay[s] River. It is the only river to fully cross the Appalachian Mountain chain from east to west (Ayers Survey, p. 3). In the opinion of Dr. Holland, it was a significant passageway in pre-historic times (Holland Affidavit).

To show that there were reasonable grounds for the state's failure to produce this evidence before the FPC, the application charged Appalachian with keeping the surveys secret:

The available evidence suggests that, although both the Ayers and Holland Surveys were commissioned by the licensee, Appalachian Power Company, they were never transmitted to the FPC and the FPC, after requiring them, never asked for their results.

As far as Petitioners have been able to determine, the Surveys are not a part of the record of the FPC in this case. Mr. Paul Johnson, who was project engineer for Appalachian, stated that the Surveys were never transmitted to the FPC and were kept secret (Poole Affidavit). The U.S. Department of the Interior admitted never having made the survey provided for under 16 U.S.C. Section 469a–1 (Poole Affidavit). Dr. Holland states that he furnished copies of his survey only to Appalachian Power Company (Holland Affidavit).

North Carolina regarded this failure by Appalachian and the FPC to consider or even disclose what archaeological and historical resources would be affected by Blue Ridge as another example of their disregard of the true costs of licensing the power project. In 1971, when Appalachian was in possession of all the data from both archaeological surveys, the company had filed a document entitled Applicant's Environmental Assessment for the Blue Ridge Project. It contained only the following information with regard to historic and archaeological sites:

Preservation of significant and historic sites.—While promoting the economy of the region through the construction of the Blue Ridge

Project, Applicant [Appalachian Power Company] also intends to co-operate with all appropriate public agencies in preserving the project area's existing values. In this connection, Applicant will cooperate with such agencies in the identification of any structures of historical signifi-cance, which may be located within the area to be inundated by the reservoirs, with a view to their possible relocation. Applicant will report to the Federal Power Commission if any of these agencies believe that historical considerations necessitate the relocation of structures so identi-fied and will make available such reasonable funds as the Commission, after notice and opportunity for hearing, may approve or direct for the purpose of effecting such relocation. . . . Also, Applicant will cooperate with the Smithsonian Institution in the identification and salvage of articles of archaeological significance, if any, located in the areas to be covered by the project reservoirs.

North Carolina thought this document misleading in the light of what Appalachian knew at the time. Moreover, the FPC never pressed Appalachian for the data, although the surveys were carried out at its request. The FPC's environmental impact statement con-tained no information whatsoever about any historical or archaeo-logical sites within the project area. Under a 1974 amendment to the Historic Preservation Act of 1960, any federal agency, if it finds or is notified that a federally licensed project "may cause irreparable loss or destruction of significant scientific, prehistorical, historical or archaeological data," is required to notify the Secretary of the In-terior in writing so that a survey can be conducted and action taken to preserve, protect, or recover the material or structures. Because the FPC had not investigated the possible existence of archaeologi-cal data, Interior was never notified to undertake a survey of either the historical sites or the archaeological sites within the project area.

The FPC's answer to North Carolina's charges was that including the surveys in the record would not have made any difference; it still would have issued the license for Blue Ridge. It further said that there would be ample time while Blue Ridge was being con-structed to carry out the necessary historical and archaeological surveys and recovery. The Court of Appeals, in response to the state's application, on October 9 ordered that the matter be de-ferred pending the oral argument on the case on October 23.

The day before the oral argument, North Carolina received more good news. The Department of the Interior informed state officials that the Bureau of Outdoor Recreation had officially accepted the state's management plan for the scenic river and would start work on a draft environmental impact statement. This would be a job of several months; but after it was done, the proposal to declare the New a national scenic river could be circulated for comment, and the ninety-day review period would begin. James Watt, the director of the bureau, who had been nominated to a vacant seat on the FPC, had withdrawn from consideration of North Carolina's application to avoid a possible conflict of interest, thus removing a cause for delay in consideration of the North Carolina plan. The bureau acted despite pressure from Virginia officials. Governor Godwin, in a letter to Secretary Kleppe, had "urged strongly" that the department defer circulating North Carolina's application until the Court of Appeals ruled on the validity of the license. Congressman Caldwell Butler of Virginia also wrote Kleppe that he was "deeply concerned about the effect of action at this time by the Department of the Interior on the litigation now in process before the Court of Appeals." He urged Kleppe "to delay any such action until this matter is resolved [by the court]." Secretary Kleppe remained noncommittal on his ultimate decision, but he allowed things to proceed.

The oral argument on October 23 was held at the U.S. Court of Appeals in Washington, D.C., before Chief Judge David Bazelon and Associate Judges Spottswood Robinson III and Roger Robb. The judges seemed receptive to North Carolina's arguments that the FPC had failed to consider peak-load pricing or scenic-river designation. They also directed hostile questions at FPC attorney Steven Taube as to why the agency had ignored the archaeological data. The judges expressed just one concern about North Carolina's case, asking whether the state's petition for rehearing, filed with the FPC after the license was granted in 1974, raised the peak-load pricing and scenic-river issues. They stated that even though they might agree with North Carolina on the merits of these issues, the court might lack jurisdiction to consider them if they had not been properly brought to the attention of the FPC. Smith, Rich, and I, as attorneys for the state, argued that although peak-load pricing was not specifically mentioned in the petition for rehearing, the FPC *was* asked to reevaluate the power project on the ground that conservation measures might make it unnecessary. Likewise, the scenic-river alternative was not brought up specifically in the petition, but the need to study all alternatives under the National Environmental Policy Act was generally alleged. This, it was argued, was sufficient since it was the FPC's job, not North Carolina's, to consider and study these problems.

Despite this difficulty, the North Carolina forces were jubilant at the conclusion of the argument. Appalachian and the FPC had been pressed extremely hard by the court. Steven Taube was overheard telling an associate, "I think we'll be remanded." Attorney General Edmisten, who attended the argument, told reporters that there would be a remand by the first of the year.

The supporters of scenic-river status for the New now believed, somewhat prematurely, that only Secretary Kleppe remained as the last hurdle to be overcome. They decided to step up their political campaign in preparation for the final assault. Meeting at Hawks Nest State Park in West Virginia on October 25, the Committee for the New decided to mount a national campaign to save the river. The issue had been given increasing attention in newspapers and other news media all over the country. Wallace Carroll, former publisher of the *Winston-Salem Journal* and a director of the committee, had facilitated this publicity by calls and letters to his contacts in the newspaper world. People in many different states had been in touch with the Committee for the New, and the members decided to capitalize on this by renaming the organization the National Committee for the New River and organizing chapters in many different states. In addition, another Festival of the New was planned for January in order to draw attention to the issue. New information about the effort to preserve the river was prepared and mailed to committee members in an effort to encourage them to send letters and petitions to Interior.

In November Kleppe received political pressure from both sides in the fight. On November 5 Congressman Neal met with Kleppe to ask him for speedy action on the scenic-river application. The Committee for the New wrote Kleppe that he had a legal obligation to circulate the application and suggested that they would go to court to compel him if he delayed much longer. Both governors, Holshouser and Godwin, also met with Kleppe to argue their respective positions on the matter.

On November 26, 1975, the day before Thanksgiving, Kleppe announced that on November 28 he would begin circulating North Carolina's application together with a draft environmental impact statement to other federal agencies, thus beginning the ninety-day review period required by statute. It was a rebuff to Godwin and a victory for Holshouser. Kleppe was the fourth Secretary of the Interior whom Holshouser had had to persuade to accept the state's application.

Kleppe, however, refused to commit himself on how he would rule after the ninety-day period or to say whether he would even make a decision if the Court of Appeals had not acted by that time. He further stated that if the court affirmed the grant of the license, the scenic-river designation would have no effect. In spite of the uncertainty, it was a good Thanksgiving Day in North Carolina.

The opening of the review and comment period was regarded by both sides in the dispute as a declaration of all-out war. Each faction fought with the weapons it had at hand. This time, in contrast to 1974, the principal forces behind the preservation of the New, the State of North Carolina and the National Committee for the New River, were ready. By encouraging the organization of local affiliated chapters in areas all over the country, the National Committee was able to generate a multitude of letters and petitions to the agencies and to Congress on behalf of the scenic-river application. National conservation organizations such as the Sierra Club, Audubon Society, Izaak Walton League, and others cooperated by featuring the New River issue in their publications. Secretary Kleppe received an average of fifty letters a day against the power project from people all across the country.

The national news media were soon following the New River

issue. By January, 1976, over one hundred fifty newspapers from every section of the country had written editorials in favor of North Carolina's scenic-river plan. Probably most of the editorial writers had never seen the river, but there was great appeal in the story of a bunch of mountaineer farmers struggling against a big power company. The primary reason for the media blitz, however, was that Wallace Carroll was busy writing and sending the National Committee's material to everyone he knew in the newspaper business. For example, the following editorial appeared in the *Daily Hampshire Gazette* (Northampton, Massachusetts) on December 19, 1975:

An editor for whom we once worked down there and whom we respect has alerted us to what is without question one of the most reckless and devastating acts in the nation's headlong quest for energy at any cost.

A huge power company plans to dam up the New River in North Carolina, flood 40,000 acres of rich bottomland, pasture and forest, displace 3,000 rural folks living on the land—all to generate power to meet peak energy demands in far-off communities. The calculations of people who oppose the damming of the New River are that the power project would actually burn up more energy than it produces at a time when the nation has an energy shortage.

It sounds crazy but it will happen unless within the next several weeks the new Secretary of the Interior Thomas S. Kleppe decides to accept this 25-mile stretch of the New River into the national scenic river system. Our editor friend believes that if the New River . . . is taken into the national scenic river system, the power project would be killed. He says this will not happen, however, unless Kleppe and President Ford are aware of the strong sentiment in all parts of the nation for protecting the New River.

In addition to the daily press, support for the New River appeared in other media as well. Dan Rather of CBS News featured the river on one of his television news programs; syndicated columnists of every political stripe from Jack Anderson to George Will devoted articles to saving the river. *Newsweek* ran an article entitled "Of Time and the River" that generally favored the preservation of the New.

Some of the people who responded to the campaign had grown up in the area of the power project but had left and were living in other parts of the United States. At his home in Oklahoma, fifty-four-year-old Paul Barker recalled his childhood in Ashe County in a poem called "New River":

> In Ashe County so far away,
> In a place I call 'back home';
> To my friends still there, and newcomers too,
> I dedicate this poem.
>
> I was born in a cabin on Silas Creek,
> At night we lit an oil lamp.
> We didn't have any plumbing;
> And man, that outhouse was damp!
>
> I often wonder if that one room school
> Still stands beside the road,
> Or if it gave way to progress
> Because of some building code.
>
> I caught my first bass in New River,
> A good memory of the past
> The water was cold from the mountains,
> Running free and clear and fast.
>
> I took nature's beauty for granted,
> Never dreaming a proposal would come;
> To deface and bury nature's wonder,
> For money and power for some.
>
> It happened on TV news one night,
> I was twelve hundred miles away.
> They talked about damming New River,
> For power and progress, they say.
>
> Statistics were given by newsmen
> Of the dollars the dam would cost,
> But little was said of the history,
> Or man's heritage that would be lost.
>
> We now know that early man
> Inhabited our mountain land.

So why should we allow our history
To be buried by a power company dam?

The dam people care nothing of history;
Their's is a monetary scheme.
So we must make them leave it alone
And not harness that beautiful stream.

We must keep the land as it is.
Keep it unspoiled by those few
Who are interested only in dollars,
And not caring what their 'progress' will do.

The nation is watching us my friends;
Let's keep our mountains unharmed.
We must preserve heritage through the years
And not caring what their 'progress' will do.

John Baldwin, eighty-two years old and ailing, wrote a poem called "Run River Run" from his hospital bed in Beckley, West Virginia:

John Milton, a blind poet,
No . . . that's not me,
But if he'd had eyes to see
He would have loved New River.
Piney Creek is a New River arm,
And there we used to make our farm . . .
Long walks to school, gentle folk, sweat,
Tears, love and laughter . . .
You know what we were after: Peace!
And from the Almighty Giver
It came in small doses . . . mostly by New River.
Then came the war, when boys and men became the same:
At Folkstone and Bellicourt and by the River Seine.
Back home again to Caroline,
My mother's name . . . my state so fine.
New River, adventurous and beautiful,
Led north to source of fortune; coal mine.
This life, this land's been good to me.

And now I find it hard to see
Why it's such a necessity . . . to dam New River.
Oh, I know that memories must fade and die
Just the same as you and I . . .
But if you pray to God above
To thank him for his blessed love . . .
Then ask him, PLEASE, to
Damn the dam . . . but not New River!

Appalachian Power Company officials were powerless to match the National Committee's media blitz. Only a few newspapers in Virginia supported them editorially, and most of the publicity generated in favor of the power project consisted of letters by and interviews with power-company officials. Appalachian did, however, enjoy easy access to high federal officials, many of whom were development-minded. Joe Dowd and members of the Virginia congressional delegation quietly made the rounds of high officials of the Army Corps of Engineers, Department of Agriculture, Federal Energy Administration, and Interior and argued that the opponents of Blue Ridge were distorting the facts and were guilty of demagoguery in their campaign against the project. Something more was needed, however, to counteract the national media campaign.

American Electric Power Company (AEP), Appalachian's parent company, decided to try a media blitz of its own—a national advertising campaign. They had used this weapon before, in 1974, under Chairman of the Board Donald Cook's leadership. At that time AEP had spent about $3 million on full-page ads in newspapers and magazines such as the *New York Times*, the *Wall Street Journal*, *Time*, and *Newsweek* to attack the controls on sulfur-dioxide emissions in the Clean Air Act. That campaign had been successful; maybe a similar one would work on the New River.

In the first few days of February, 1976, full-page ads entitled "The Truth about The Blue Ridge Project" began appearing in newspapers all over the country. In these messages, AEP extolled the merits of the power project and attacked the suitability of the New River as a national scenic river. The press was attacked as prejudiced, and those in North Carolina who opposed the project were portrayed as "selfish elitists." AEP told the story from its perspective:

The welfare of this entire nation is endangered by an energy shortage. The White House has ordered the development of all our energy resources—and Departments of Government are trying to do just that, in keeping with environmental standards. No selfish group which stands in the way can remain unchallenged—be they privileged elitists or a prejudiced press.

.   .   .   .

Blue Ridge:
—will conserve our national resources by consuming no oil and gas.
—will provide emergency reserve power for the East Central region of the United States.
—will provide 160,000 acre-feet of flood control capacity—endorsed by the U.S. Corps of Engineers—where none exists today.
—will assure water benefits downstream where, to improve recreation and fishing, the river flow is periodically in need of augmentation.
—will vastly increase the recreational potential of the area, turning it into one of the most appealing sites in the East.
—will facilitate the economic development of depressed Appalachia.
—will consume less fuel than any available alternative means of generation.

.   .   .   .

And so, a license was granted—effective January 2, 1975.
The sum of the benefits to the American people was so demonstrably great that the Blue Ridge Project won the support of the States of Virginia and West Virginia, the Federal Power Commission, the Federal Energy Administration and for six long years ('67 to '73)—until a mysterious reversal—the State of North Carolina.
Many North Carolinians fully favor the project.
But, it is not welcomed by an affluent few. They shudder at the thought of intrusion by outsiders.
They have decided to resist the needs of this nation . . . to ignore the President's call . . . and to save the privileged status quo by killing the Blue Ridge Project.

Twice they tried in the U.S. Congress. Once with a rider on the Rivers and Harbors Bill.

They failed.

Once they actually tried to have this tame, this bridged and dammed river-along-the-highway made a component of the untouchable National Wild and Scenic Rivers System . . . a flagrant perversion of an Act of Congress. They failed.

Ironically, not one word of criticism of these actions appeared in the press.

And now . . . THE STING!

Although the people of North Carolina will benefit substantially from a strengthened power supply, our gift of 3,900 acres hand-picked by North Carolina for a lake-front State park, recreational facilities valued in the millions, and participation in a construction payroll of over $125,000,000 . . . the influential elitists are about to euchre them out of it with a tricky scheme.

Incredibly, North Carolina officials would circumvent the U.S. Congress by having a *limited stretch* of the New River incorporated into the National Wild and Scenic Rivers System . . . by administrative decree.

Just enough of a stretch to block Blue Ridge!

This campaign largely backfired, however. The farmers and other people of New River Valley joked about being called elitists by the largest power company in the United States. The charge ignored the fact that by this time the movement to preserve the New was truly broad-based; letters were pouring in to Interior and the other agencies from all around the country. The advertising campaign sparked a new round of anti–Blue Ridge editorials, especially in North Carolina. People in North Carolina bristled. If they had ignored the New River issue up to now, they could no longer remain indifferent. The matter was becoming an important political issue in the state. It could not be overlooked that President Ford and Ronald Reagan in the spring of 1976 were engaged in a close struggle for the Republican nomination for president. And one of the key early primaries would be held in North Carolina on March 23, 1976. Political advisers for both candidates were closely watching the North Carolina situation.

Despite all this lobbying and media activity, the federal-agency and other comments that were filed with Interior before the close of the ninety-day period on February 27 divided along fairly predictable lines. The FPC, American Electric Power, and Virginia strongly criticized the scenic-river plan as well as the environmental impact statement. Their comments attacked the impact statement as a violation of the National Environmental Policy Act, since Interior, it was alleged, had failed to discuss the environmental costs and benefits of all alternative courses of action. They contended the Blue Ridge Project would be environmentally more beneficial. It was obvious that they were preparing to file suit against Interior to challenge the agency's decision in court if it decided in favor of North Carolina. On the other hand, the comments of North Carolina, the Environmental Protection Agency, and several conservation organizations strongly favored the scenic-river designation.

Notable by their absence, however, were any strong comments one way or the other by federal agencies directly under the control of the executive branch of the government and the administration of President Ford. The Army Corps of Engineers was noncommittal, and Agriculture did not comment at all. Another agency that had nothing to say was the Federal Energy Administration (FEA).

Behind the scenes, however, a massive struggle had taken place. In October, 1975, Governor Holshouser had met with Frank Zarb, FEA's administrator, and was convinced he had obtained Zarb's commitment to oppose Blue Ridge. In the succeeding months, however, Zarb signed letters prepared by his staff favoring Blue Ridge. When Interior's environmental impact statement on the scenic river came to the agency, it precipitated an internal struggle over policy. The Energy Research and Development Office strongly favored Blue Ridge, while the Environmental Programs Office opposed it. Both prepared draft comments, reflecting their different points of view, for Zarb's signature. Zarb's deputy, John Hill, called both groups together and gave them five days to prepare an extensive analysis of their competing positions. Ellen Brown and other employees of the Environmental Programs Office worked day and night composing a critical review of Blue Ridge. The Energy Research and Development Office drew its review primarily from the FPC's environmental impact statement. Zarb and Hill, after reviewing this information, backed down and held the FEA to a neutral position on the Blue Ridge Project. It was also rumored within the agency that the White House had quietly contacted Zarb and told him to lie low.

Once the review period had expired and the comments had been collected, it was up to Interior to make a decision on the matter. Kleppe had still not announced what position he would take: the prognosis was that the decision, even if favorable, would take several more months, since NEPA required that a final environmental impact statement be prepared and circulated together with Interior's final proposal. The Court of Appeals' decision would be handed down shortly. As Kleppe's staff argued that he should not make his decision in advance of the court's, he agreed to put it off until April. In normal times Kleppe would undoubtedly have held to this schedule, but 1976 was far from normal. Presidential politics intervened to aid the effort to save the New River.

Throughout the ninety-day review period, North Carolina officials were aware that the comment period would expire just before the March 23, 1976, North Carolina presidential primary. They sought to take advantage of this fact by persuading Kleppe and the President to declare themselves on the issue during the primary campaign, when the most political leverage could be brought to bear on them. Furthermore, Governor Holshouser was President Ford's Southern campaign chairman, and he made it clear to the White House that the New River issue was of paramount importance to him personally.

Holshouser wrote Ford an extensive memorandum on the New River issue in February, outlining all the alternative approaches the President could take on the issue during his campaign appearances within the state and the likely political consequences that would result from each alternative. These boiled down to three choices: supporting the scenic-river status, opposing it, or staying neutral. Holshouser advised that coming out against the river would be politically disastrous, and that staying neutral would be hard because he would be constantly questioned on his stand by reporters; thus the only reasonable course was to come out in favor of the scenic river. Further pressure was put upon Ford because, on February 6, 1976, Ronald Reagan, while campaigning in Greensboro, had come out firmly in favor of preserving the New River, stating that it would be a "disaster to destroy such a splendid stream."

By early March the President had made his decision—he would fully support North Carolina's scenic-river application and would direct Kleppe to decide in favor of the state. Another difficulty remained, however. In order to have a maximum political impact, the decision would have to be announced well in advance of the primary. But to satisfy legal requirements, a final environmental impact statement had to be prepared before any announcement. Furthermore, Interior was probably going to be sued by Appalachian and Virginia after the decision was made, and it would look unseemly if the President made the first public statement on the decision in the midst of a political campaign. Accordingly, the White House decided that Interior should embark on a crash program to prepare the final impact statement and that Kleppe, not the President, would make the public announcement as soon as it was ready.*

On Friday, March 12, 1976, the impact statement was completed and distributed, and the next day Kleppe announced his decision to sign the official order designating the New as a national scenic river. He declared that he would wait an additional thirty days, however, before actually signing the designation and publishing it in the *Federal Register*, since legal guidelines required that an agency wait at least thirty days after the submission of a final environmental impact statement before taking final action. He also

* Ironically, President Ford was destined to lose the North Carolina primary on March 23, despite his New River decision. Apparently, in the public mind, the decision was thought to be Kleppe's, not Ford's.

stated that Interior's legal advisers felt that if the Court of Appeals upheld the FPC's grant of the license, the scenic-river designation would have no effect.

This last qualification was lost, however, in the general jubilation that greeted Kleppe's decision. The news media announced that the New was saved. The National Committee for the New thought the job was over. On March 15 Congressman Ken Hechler of West Virginia called for a victory celebration to be held on the banks of the river.

The rejoicing was short-lived. On March 24 the Court of Appeals in Washington finally handed down its long-awaited decision. It upheld the FPC's grant of the license, revoked the stay order, and said the FPC could declare the Blue Ridge license immediately effective. The basic ground for the court's determination was exactly what I had feared when I began to work on the case in the fall of 1974: the judges refused to consider North Carolina's main contentions because they had not been properly raised in the Petition for Rehearing filed with the FPC. Thus the agency in licensing Blue Ridge was free to ignore the scenic-river controversy and the peak-load pricing of electricity. As for the human costs of the power project and the disruption of life in the New River Valley, the court merely said that those issues had been thoroughly considered by the FPC and adequate relocation assistance would be provided. The court dealt with the question of the newly discovered archaeological data by ordering the FPC to require Appalachian to provide the necessary time and funding for "complete excavation and salvage."

The court's decision on Governor Holshouser's scenic-river application to Interior was particularly disappointing to North Carolina. It ruled that the application was validly made: a governor can ask federal scenic-river status for a river that has not been recommended for this designation by Congress. But the river involved is not protected against an FPC license during the period it is under study by Interior. Thus, before final designation, the FPC could license the power project.

This decision shocked the people of the New River Valley. After years of despairing that Blue Ridge could be stopped, they had only recently come to believe they would win. Now—at the brink of success—victory had been snatched from their grasp. Many people wept; others held their anger in silence. Was there anything that could be done? Was this the end of the line? There were no quick or easy answers.

10

The New River Like It Is

AT THE TIME SECRETARY KLEPPE ANNOUNCED
his intention to designate the New River a national scenic river, he
set April 13, 1975, as the date for making this declaration. The decision of the Court of Appeals on March 24 upholding the right of
the FPC to declare the Blue Ridge license immediately effective
threw everything into confusion. Would Kleppe still sign the scenic-
river designation on April 13? Even if he did, would it be valid?

Just after the Court of Appeals' decision, I received a phone call
from Ernie Carl, who put the question of the hour, "What do we
do now?" We discussed two immediate needs. First, all the opponents of Blue Ridge and their allies had to be held together; it
was no time to give up; a battle had been lost, but not the war.
Second, a new strategy had to be devised to carry on the fight.

In the valley of the New, the balm of Gilead trees were blooming,
and it was time for spring plowing. Steve Douglas, at his farm on
the South Fork, expressed the predominant mood: "Most people
just flat aren't giving up. They're just bound and determined to
stick this thing out, to stay with it till we win. All this has made
people realize the bureaucratic system that we're up against." In
Congress, Senator Helms and Congressman Neal both announced
their intention to seek legislation to save the New River. On March
27 key members of the National Committee for the New River met
at Ham Horton's house in Winston-Salem and determined to press
for a bill in Congress to declare the 26.5-mile stretch of river from
Dog Creek to the Virginia line a national scenic river. Attorney
General Edmisten announced that he would seek immediate review
of the Court of Appeals' decision in the U.S. Supreme Court. In
faraway Alaska, Everett Newman, Jr., a Grayson County native

serving in the military, wrote an open letter to his fellow Virginia citizens urging them not to give up:

At this writing I am at Ft. Wainwright, Alaska, on a three-year active-army tour. I have just received the news that the U.S. Court of Appeals has authorized the start of construction on the Blue Ridge Project. This news lays heavy on my heart and, once again, I feel the need to add my voice to the plea to save New River in its natural state.

They have given their consent to destroy the New River Valley, to cover 40,000 acres with water, and to force 900 families to leave their homes. Who are "THEY"? Certainly not the 900 families being forced from their birthright. Certainly not the majority of the citizens of Grayson County. Not even the citizens of the area who favor the dams. "THEY" are three Washington judges who probably have never seen the New River Valley and could never hope to understand the lifestyle that flourishes there. Are these the people we want to shape our futures? I think not!

. . . .

Just what is good for the people of Grayson County? Appalachian Power Co. would have us believe that we will be swept from certain poverty and ruin to fabulous wealth and riches. They call it progress . . . I call it BULL! There may be some minor jobs open to the local populace, but the main construction will be done by strong labor union forces. Work forces from another area who will rape our land and then scurry back to their suburban sanctuaries far away and leave the Grayson County citizens with their broken dreams.

. . . .

Have we become so blinded by greed that we have refused to use common sense? The project will use more energy than it produces and, yet, we are allowing it to be built in the name of progress. This is absurd and inexcusable.

I urge each of you to write to our representatives in Congress and demand that they introduce legislation which would override the court's decision to uphold APCO's license . . . SAVE NEW RIVER for all of us and all the generations following us.

On March 30 I met with Norman Smith, Millard Rich, Ed Adams, John Curry, and Art Cooper at Ernie Carl's office in Raleigh to plan a new strategy for dealing with the situation. We decided to proceed on a broad front, in all three branches of government and before the FPC. This technique of pursuing several alternatives at the same time gave North Carolina multiple possibilities for success, since any one of them could be decisive in the struggle.

First, we decided to ask the Department of the Interior to formally designate the New a national scenic river on April 13 as scheduled. The state could then post signs along the river and begin to implement the management plan. North Carolina would then go into court to protect the national scenic river from the FPC and Appalachian. The vehicle for this action would be the pending appeal in the Court of Appeals in Richmond, which had been stayed since August. Immediately after Interior's designation, North Carolina would file a motion in that suit asking for leave to file an amendment to the complaint, stating that under the Wild and Scenic Rivers Act, the FPC is specifically prohibited from taking any action to adversely affect a national scenic river. This action, we hoped, would prevent the FPC from declaring the Blue Ridge license effective after the final designation had taken place.

Second, the state, through its congressional delegation and the National Committee for the New, would seek legislation in Congress to have the New declared a scenic river. We decided that congressional action should be consistent with the state's view that, when Interior's formal designation came, the New would be a national scenic river. The bills to be introduced in Congress, then, would merely affirm Interior's designation and prohibit the FPC from disregarding that designation.

Third, North Carolina had to do everything possible to prevent the FPC from declaring the license effective before Interior's designation, which could not be made until April 13 at the earliest. We decided to file a motion with the FPC to delay any move to declare the license immediately effective. We cited three reasons why the FPC should not declare the license immediately effective: (1) the archaeological data had been kept secret by Appalachian, and the FPC should permit formal hearings on the extent of the archaeological remains; (2) Secretary Kleppe would shortly declare the New a national scenic river; and (3) the State of North Carolina intended to obtain Supreme Court review of the Court of Appeals' decision. This motion was filed with the FPC on March 31.

Fourth, we decided that North Carolina should file a petition for a writ of certiorari with the U.S. Supreme Court, formally asking for a reversal of the Court of Appeals' decision. The filing of this petition would extend the period of judicial review of the FPC's license for Blue Ridge.

But while we were working, Appalachian and the FPC were busy too. They also knew that it was important whether the license was declared effective before Interior formally signed the scenic-river designation on April 13. They resolved to act immediately, before anything else could happen to force a delay. On March 24, the same day the decision was handed down, the four FPC commissioners met to complete action on Blue Ridge. They directed the staff to draw up an order modifying the license as the court had directed and de-claring it immediately effective.

This order was formally issued by the FPC on March 26. It modi-fied Article 33 of the Blue Ridge license, requiring Appalachian to "provide the necessary funding for complete research, excavation and salvage, all of which must be completed prior to construction and/or flooding, whichever is applicable." The order then stated: "The license for Project No. 2317 [Blue Ridge] is in effect."

The FPC did not give anyone in North Carolina notice of this order, either before or after it was entered. Lawyers for the state found out about it on April 2, when Washington-based reporters for several North Carolina newspapers heard about it and relayed the news. We were furious. Our March 31 motion asking the FPC to delay the effective date for Blue Ridge and to hold hearings on the archaeological data was moot even before it was filed.

Appalachian was jubilant; it had its license at last. The FPC regarded its job as done. Kenneth Plumb, the secretary of the FPC, said, "The case is all over now as far as the commissioners are concerned."

In North Carolina it was time for some fast action. The FPC's order could not be allowed to stand. The New River itself was physically threatened; there was nothing to stop Appalachian from immediately beginning earth-moving and construction activities. Furthermore, the order threatened North Carolina's whole strategy. Interior might rethink its position; the case in the Richmond court might not be viable if the license became effective before Interior acted; the Supreme Court would not act for over a year, by which time there could be a dam on the New River; and Congress might hesitate to act as well. The state needed another stay to hold up the project. The logical place to go would be to the Supreme Court, where it might be possible to obtain a stay from the Supreme Court justice assigned to the District of Columbia circuit. This, however, was Chief Justice Warren Burger, whom we regarded as unfriendly to the state's position.

There was another way to get a stay of the project, however. Norman Smith pointed out that Rule 41 of the Federal Rules of Appellate Procedure provides: "The mandate of the court shall issue 21 days after the entry of judgment unless the time is enlarged or shortened by order."

As a matter of law, once an appeal is taken, jurisdiction of a case, known as the *mandate,* is vested in the Court of Appeals, and the authority of an administrative agency is suspended so that it cannot proceed with a case until the mandate has been returned to it by the court. In the case of Blue Ridge, the FPC had attempted to act before the mandate had been returned. Thus, Smith reasoned, under Rule 41, the FPC's order of March 26 was void and illegal because the commission did not have jurisdiction! Furthermore, the commission would not be able to act until twenty-one days after the Court of Appeals' decision; this would not be until April 14, one day *after* Interior was due to formally sign the scenic-river designation for the New!

Accordingly, we prepared to attack the validity of the FPC's order of March 26. On April 4 we filed a petition for rehearing with the FPC and a motion that the commission rescind its March 26 order. Under the Federal Power Act, if the FPC refused this motion, North

Carolina could immediately appeal the refusal and begin yet another lawsuit against the commission, asking the Court of Appeals to declare the March 26 order illegal.

There was still another possibility for a stay. On April 8 North Carolina filed a motion for stay of mandate with the Court of Appeals in Washington. The state argued that it intended to file a petition for a writ of certiorari with the Supreme Court and that substantial grounds existed for questioning the legal validity of the court's March 24 decision. North Carolina furthermore argued that the status quo should be preserved and the license not be allowed to go into effect because the Secretary of the Interior would shortly designate the New a federal scenic river and bills were pending in Congress to confirm Interior's designation.

Appalachian and the FPC tried mightily to prevent the issuance of another stay. They argued that the application for review in the Supreme Court would be a frivolous move solely for the purpose of delay. Appalachian stated that while it had no immediate plans for construction, it did want an effective license.

Smith and I then went to work drafting the state's petition for a writ of certiorari in the hope that the Court of Appeals would grant the motion to stay the mandate. We waited and hoped. The month of April went by and still the Court of Appeals had not acted; but they had withheld the mandate while the judges considered the state's motion. Finally, on May 5, 1976, Judges Bazelon, Robinson, and Robb rendered their decision. This time they ruled in favor of North Carolina:

On consideration of petitioner's motion for stay of mandate and of the opposition thereto, it is
ORDERED by the Court that petitioner's aforesaid motion is granted and the Clerk is directed to stay the issuance of the certified copy of this Court's judgment to and including May 14, 1976.

This meant that, under the Federal Rules of Appellate Procedure, North Carolina could obtain a stay of the Blue Ridge Project and the effective date of the license by filing its petition for writ of certiorari in the Supreme Court by May 14. We prepared the petition and filed it with the Supreme Court. Once again, time had been gained to mount another effort to save the New River. The Supreme Court would not act on the petition until sometime in the fall.

Meanwhile, the state had been extremely successful in its effort to keep the Department of the Interior on its side. On April 13, as scheduled, Secretary Kleppe signed the scenic-river designation for the New and published it in the *Federal Register*. Attorney General Edmisten personally waited all day in Kleppe's outer office on the appointed day to make sure the Secretary would act. Kleppe also promised to ask the U.S. Department of Justice to file an *amicus* brief in the U.S. Supreme Court supporting North Carolina's petition for writ of certiorari. He stated that he was taking the action "to protect the integrity of my decision." Promising that Interior would support North Carolina "the rest of the way," he declared: "I want to announce here and now that the President is very much in support of these actions."

Interior also carried out, on April 7 and 8, a preliminary survey of the archaeological, cultural, and historical resources of the New River Valley. A team from the National Register of Historic Sites in the company of North Carolina officials uncovered a rich lode of historically interesting structures. Here is an excerpt from their report:

The economy of the region developed around small homesteads established on the river and the numerous creeks which form its tributaries. The first trails were along these streams. In addition, the river and its tributaries provided a source of power for the grist and textile mills which were established very early in the area. The development of water powered carding machines, in particular, encouraged the establishment of textile mills. Two of the most historically significant properties identified during the field survey were the Field's Manufacturing Company, an 1884 Woolen Mill on Wilson Creek in the little town of Mouth of Wilson, Virginia and the grist mill on Dog Creek. The Field's Manufacturing Company . . . which would be completely destroyed by the Blue Ridge project, is, according to the Historic American Engineering Record and Tom Leavitt of the Merrimack Valley Textile Museum, an extremely rare, perhaps unique, industrial complex in which some of the old equipment, including at least one ca. 1884 carding machine originally run by water power, remains. The scale of operation, age of equipment, original site and continuation of production of woolen textiles as well as its longevity of operation make it likely that this mill would qualify as a National Historic Landmark. Its inundation would be a significant loss. The mill's value is also enhanced by its location in the unincorporated village of Mouth of Wilson, where life historically has revolved around the mill, as it does today. Cochran Grist Mill on Dog Creek is a 19th century water driven mill with its ca. 1870/1890 Fitz Wheel still intact, one of about a half dozen of such wheels recorded by the Historic American Engineering Record. It has also retained several mill-related outbuildings and is representative of the many grist mills which once operated on the New River and its tributaries.

.   .   .   .

The preliminary survey indicates that the New River area with its 19th century farm complexes, Victorian farmhouses, mills, rolling farmlands, and small communities, is particularly significant because of the high degree of historical integrity which it has maintained. Due to its isolation and low population, there are very few 20th century intrusions.

Within a week after this preliminary survey, the State of North Carolina put a twenty-man team of archaeological, historical, and architectural experts in the field to make an in-depth study of the cultural resources of the New River Valley. They were directed by Dr. Larry Tise, director of the North Carolina Division of Archives and History, and Stephen Gluckman, the state archaeologist. This team, together with several volunteers and local residents, put in sixteen-hour days, seven days a week, working in the valley. Ben and Linda Robertson, archaeologists from Brown University who developed the methodology for the study, were surprised by the archaeological richness of the area. They discovered over 160 Indian sites, ranging from small encampments to large villages, and stone implements dating from 8000 B.C. down to historic times. Several of the sites discovered were characterized by stratified soil layers, leading the Robertsons to conclude that excavation of these areas could add significantly to what is known about prehistoric man in the eastern United States. Michael Southern, the architec-

Grassy Creek United Methodist Church, Ashe County. (*Courtesy of Department of Archives and History, Raleigh, N.C.*)

Cochran Gristmill, a 19th-century water-driven mill on Dog Creek in Ashe County. (*Courtesy of Department of Archives and History, Raleigh, N.C.*)

The Gambill-Crouse cemetery, Alleghany County. (*Courtesy of Department of Archives and History, Raleigh, N.C.*)

The William Weaver house, Alleghany County. (*Courtesy of Department of Archives and History, Raleigh, N.C.*)

tural historian on the expedition, was equally enthusiastic about the many nineteenth-century buildings in the valley. He found and studied two antebellum brick houses and a multitude of frame, Victorian-style residences with two-tiered porches and impressive woodwork in cornices and eaves. In many areas, he found that farm complexes and crossroad communities exist little changed in outward appearance since the late nineteenth century. As a result of this survey, on May 3 Tise nominated thirty-three historical, architectural, and archaeological sites for inclusion in the National Register of Historic Places and announced that more sites were under study for possible nomination.

During the spring of 1976, the State of North Carolina was sparring with the FPC and Appalachian in the Court of Appeals in Richmond. In April, after the Washington, D.C., Court of Appeals' decision of March 24, the FPC moved in the Richmond court to dismiss North Carolina's pending appeal on the ground that the case had become moot, since the Court of Appeals in Washington had interpreted the Wild and Scenic Rivers Act as not protecting a state-designated scenic river while it was under study by the Department of the Interior. After Interior had formally designated the New as a scenic river on April 13, North Carolina countered with a motion asking the court to take jurisdiction and to grant the state leave to file an amendment to the complaint, which would allege that the New River was now not just a study river, but a formally designated national scenic river and thus granted all the protection of the act. If this argument were accepted by the court, the Blue Ridge license would never become effective and the power project would be blocked. On May 17, 1976, the Richmond court reserved judgment on both the motion to dismiss and the motion for leave to amend the complaint and called for briefs and oral argument on the issues. The case was scheduled for consideration in the fall.

While this flurry of legal activity gave North Carolina and the National Committee for the New two separate chances to win the preservation of the New River in court, it also provided valuable time so that a more direct campaign could be mounted in Congress. Secretary Kleppe's action of formally designating the New a national scenic river had created a situation that was perhaps unique in American history. Two federal agencies, acting pursuant to authority granted them by Congress, had made diametrically opposed and inconsistent decisions regarding the same issue. What is more, both agencies, under the law, had absolutely final authority in the matter. Under the Federal Power Act, the FPC has ultimate decision-making power over hydroelectric project licensing; under the Wild and Scenic Rivers Act, the Department of the Interior has final authority to act on a state application and designate a national scenic river. There had to be a way to break the impasse. The judicial branch of the government could be utilized for this purpose, but courts are very reluctant to handle such issues. It would be more direct and certain if the Congress, which had passed the two laws in question, would act to solve the problem. Congressional action on any subject, however, is a slow and difficult process. The New River was no exception.

This phase of the effort in Congress to save the New began immediately after the Court of Appeals' decision of March 24. Congressman Neal and Senator Helms introduced bills in the House and the Senate to have Congress designate the 26.5-mile stretch of the New as a national scenic river and to revoke the FPC's license for Blue Ridge. The National Committee for the New geared up for an immediate lobbying effort, and a Washington-based environmental organization, the American Rivers Conservation Council, headed by William Painter, made plans to aid the fight.

The bills did not immediately begin to move, however. In the House, the matter was referred to the Subcommittee on National Parks and Recreation. The subcommittee chairman, Roy Taylor, who coincidentally represented a western North Carolina district, refused to schedule early hearings. This concerned Neal, and there was criticism of Taylor in the North Carolina press. Taylor, however, knew what he was doing. A veteran legislator in his last term in the House, he sensed that the time had not come to start shooting from the hip. He was shrewdly biding his time and was quietly working to prepare the way for the hearings before announcing them publicly.

The situation was more serious in the Senate. The legendary Sam Ervin, who had so effectively pushed the New River bill in the Senate in 1974, had retired, and his seat was taken by Robert Morgan, a Democrat who had been the attorney general of North Carolina until 1974. Helms's New River bill stood little chance of passage without Morgan's support. Senator Helms wanted quick action and asked the Senate Interior Committee to expedite consideration of the measure.

On April 1 Morgan announced where he stood on the issue. He shocked the proponents of the bill by requesting the Interior Committee to delay consideration, saying that he wanted time to study some "unanswered questions" about the situation. Morgan asked what the effect would be on landowners along the river if the New was placed in the scenic-rivers system. "Will it deprive them of the free use of their lands and will they be compensated?" Furthermore, if the dams were not built, he wondered, "Where are we going to get the power?" According to Morgan, the environmentalists called the project a "white elephant," while the FPC claimed it was a necessity. He needed more time to consider all the arguments.

Morgan's action was immediately greeted by a barrage of adverse editorial comments by newspapers all across North Carolina. The Raleigh *News and Observer* called his remarks "baffling" and asked that he "catch up on his homework in a hurry and join the fight on the river's behalf." Other comments were less kind. It was pointed out that Morgan should know more than practically anyone else about the issue, since he had been attorney general during all the early licensing hearings on Blue Ridge, when his office had formally represented North Carolina before the FPC. The *Charlotte Observer* commented:

Sen. Morgan was not held captive by gypsies in Rumania during these 10 years of fighting over the dams. He opposed the dams himself when he was North Carolina attorney general. . . . If he doesn't join his fellow North Carolina members of Congress in fighting to stop the dams, let Appalachian Power Company name one of them the Robert Morgan Memorial Dam. That way, North Carolinians will know where he stood when a Virginia power company wanted to destroy a river in North Carolina.

The National Committee for the New River called Morgan's statements "distortions of the truth." It was a bit of gallows humor that the senator was worried about the impact of the scenic river on landowners when, if the dams were built, the entire area would be flooded. The committee pointed out that "the North Carolina plan would permit the farmers to continue to raise their crops and graze their cattle just as they have always done. . . ." Under that plan, "the state would take scenic easements on some small parcels of land along the river. But this land would be farmed by the own-ers as before. The state would also acquire about 400 acres for parks. . . . In short, it is not the North Carolina plan but the New York utility that would run the farmers off their land."

Under pressure, on April 6 Morgan modified his position some-what. He stated he would support Helms's scenic-river bill but an-nounced that he would insist on an amendment requiring that either the state or the federal government compensate landowners "for any lessening of the usefulness or value of [their] land." Mor-gan stated that he believed anything less than such a guarantee would leave landowners along the river "holding the bag, with no recourse but to start a lawsuit against the government when they are denied the right to build a house overlooking the river . . . or to put up a new barn, or sell their property for other uses."

This statement caused another burst of comment. Although some defended Morgan's position, most decried it as unnecessary and said that, as a practical matter, it would kill the chances for passing the New River bill. Former Senator Ervin was one of the most outspoken of Morgan's critics, saying from his retirement home in Morganton that he was "astounded" by the senator's po-sition, which Ervin contended was too broad. He stated in a letter to Senator Helms: "As a lawyer of many years standing and a former attorney general of North Carolina, Sen. Morgan must know that under the Constitution of the United States . . . neither the federal government nor the state government can deprive citizens of the full use of their property . . . without paying them just compensation." Ervin warned that if the New River bill was not enacted, thousands of North Carolinians would be forced from their homes and Morgan's position was a threat that could guarantee flooding of the valley. Senator Helms said that he could not accept Morgan's amendment because it was open-ended. It contained no guidelines as to how, when, or under what circumstances landowners would be compensated. Morgan was adamant, however, saying, "We are going to have an amendment, or we ain't going to have no bill."

Officials of the State of North Carolina and members of the National Committee for the New were at a loss as to how to deal with Morgan's position. They could not accept the amendment because a bill requiring an open-ended federal appropriation to pay landowners would stand no chance of passage. They were puzzled as to Morgan's motives. Was he really concerned about the landowners along the river? Or was his position merely a subterfuge to stop the scenic-river bill and ensure the building of the power project?

It was up to Ernie Carl to deal with the problem. In late April Carl, with my help, devised a strategy. First of all, we reasoned, if Morgan was really just in bed with the power company, there would have to be all-out war against him. The troops could be called out, and although it would be a tough fight, there was a good chance the bill could be reported out of the Senate Interior Committee even without Morgan's approval. Acting Chairman Lee Metcalf was willing to ignore Morgan and to schedule early hearings on the measure. But it would be much easier with Morgan in the fold. He was getting a lot of political heat; perhaps he would be willing to retreat if given the chance.

We decided to smoke Morgan out. From conversations with his aides, we knew he particularly objected to the provision in the state's management plan for the scenic river which allowed the use of floodplain zoning to implement the scenic-river plan. Carl regarded this as a phoney issue. The state had never meant to rely on this as a primary means of protecting the river. It had assumed that since zoning was required under state law anyway,* whether or not the scenic river was approved, it should naturally be taken into account in any management plan for the river. The state regarded Morgan's position as particularly ironic in view of the fact that under Article 43 of the FPC's license for Blue Ridge, zoning was to be relied upon to control development around the 345-mile shoreline of the impoundment!

But this allowed the state to give Morgan a graceful way of retreating from his position. The state would ask that Morgan drop his landowner-compensation amendment in return for a public assurance that the state would not use floodplain zoning in the area of the scenic river any differently from the way such zoning would be used in any other portion of the state. It would thus appear that the state was giving up something, when in reality, it would be doing nothing it had not intended to do all along. Morgan could take the bait, however, claim a victory in his fight for landowners' rights, and save face before the press and the public. On the other hand, if he did not take this opportunity to retreat, we would know where he stood and could take steps to oppose him with the full resources of all the groups trying to save the river.

In late April, since Morgan was on a trip to the Far East, Carl entered into negotiations with Henry Poole, Morgan's legislative counsel. On April 23 Carl suggested to Poole that the state would retreat on its use of floodplain zoning if Morgan would drop his amendment. Poole, however, wanted an amendment to the state's management plan. In an April 28 letter to Carl, he wrote:

As I have pointed out to you, the State's present management plan of June 1975 (specifically on pages 28 and 32) states that "floodway regu-

---

* Floodplain zoning has been required by law in North Carolina since 1973. Local government, with the help of the state, must delineate the floodplain. The purpose of this delineation is to control floods by nonstructural means through regulating construction in flood-prone areas. Once a community has adopted floodway regulation, landowners are allowed to engage in certain uses, such as farming and forestry, but must obtain a permit for the erection of permanent structures within the floodway. This state law was passed in response to the Federal National Flood Insurance Act, which makes federal flood insurance available but requires that a local community, in order to qualify, must adopt and administer floodway regulations.

lation statutes" and "floodway regulations" would be used as the first or primary protection and control method along the New River.

This floodway zoning which would be made by Ashe and Alleghany Counties but relied on by the State as the first or primary method of protection and control, would, of course, be accomplished without compensation to the affected property owners whose property rights would be restricted. This is certainly a perversion of the intent and integrity of the legislation establishing floodway zoning. . . .

If Senator Morgan could be assured that the new plan to be submitted to the Secretary of the Interior in accordance with Senate Bill 158 will provide that the State will not use zoning as a substitute for other forms of acquisition, but rather that all property rights would be acquired by the State by acquisition in fee simple, by easement, by cooperative agreement, or by gift, then the objection which the Senator has to the passage of the bill in its present form as submitted by Senator Helms will have been removed.

The State has claimed that (1) it never intended to use zoning as a method of acquisition and (2) under the Constitution of the United States and the State of North Carolina, a property owner could not be deprived of his property without just compensation. I believe the first point will be clarified and corrected by the State's rewriting of its management plan clarifying the zoning issue.

As for the second point, the Senator is fully aware that the taking of someone's *full* rights in their property, without compensation, is indeed unconstitutional. However, in this instance the zoning of the land under floodway regulations would constitute a non-compensable taking of property rights, with the State relying heavily on such taking as outlined in the State management plan of June 1975. This is indeed allowed under both Constitutions but certainly is not a just and fair way for the State to conduct its affairs. Floodway regulations were enacted at the federal and state level for the purpose of protecting floodways, not for the purpose of property control once a river is declared to be part of the national wild and scenic river system.

It was evident that Morgan, stung by the political criticism he was receiving in the press, was ready to take the opening given to him by the state. In an April 30 statement he said:

My staff is now attempting to get an assurance from the State that, in the new management plan required under Senate Bill 158 [The New

River bill], any reliance on floodway regulations will not be used as a substitute for other forms of acquisition such as fee simple purchase, gift, cooperative agreement or scenic easement. I am hopeful, and fully expect, that the State will make such an assurance, in which case I intend to fully support Senate Bill 158.

The state's tactic appeared to be succeeding. The whole mess was boiling down to a tempest in a teapot. Carl drafted a letter to Morgan's office on May 6 to give him the necessary assurances against the use of floodplain zoning. He could not amend the state's management plan, because that would cause problems with Interior, which had approved the scenic river on the basis of the existing plan. He thus gave Morgan the following written guarantees:

1. The State will not, through any pressure or action, attempt to cause flood plain zoning to be applied any differently in the area of the scenic river than such zoning would be applied anywhere else in the State or in Ashe and Allegheny [sic] counties themselves in the absence of a scenic river. However, the State may urge the affected counties to take whatever action they are going to take in a timely manner so that scenic river acquisition can take place in the most orderly and equitable manner possible and with a maximum of local input into what is or is not compensable taking.

2. The State will not use flood plain zoning in any way which would lessen its liability in acquiring the rights to incompatible uses not hazardous to life and property.

3. The right to public access is a safe use in all areas, and if this right is acquired, it will be by gift or compensation.

Carl hoped these carefully qualified concessions would do the trick. At the same time, Morgan was being bombarded by letters from members of the National Committee for the New asking him to drop his amendment. Newspaper comment about his position continued to be extremely harsh. Cartoonists in the state were having a field day at Morgan's expense. For example, on May 2 Dwane Powell, in the Raleigh *News and Observer*, drew a New River landowner sitting in his house and wearing scuba gear, since the waters of the Blue Ridge Project covered the house up to the chimney. He was busy writing a letter: "Dear Senator Morgan, about

Cartoon depicting a possible result of the Blue Ridge Project. (*Courtesy of Larry Barton*)

A cartoonist's view of Morgan's stand on the scenic-river bill. (*Courtesy of Dwane Powell*)

your concern over restricted use of our property under the New River legislation . . ."

Morgan made no immediate response to Carl's May 6 letter. He was upset about what he referred to as "shabby treatment" in the press. He either refused to see reporters or greeted their questions with a terse "No comment." His aides were instructed not to say anything about the matter. What his ultimate position would be on the New River bill was a complete mystery.

Meanwhile, things were going quite well for the scenic-river bill on the other side of Capitol Hill, in the House of Representatives. Roy Taylor, chairman of the Subcommittee on National Parks and Recreation, quietly lined up support among the subcommittee members and called a hearing on the bill for May 6. Secretary Kleppe wrote Taylor to declare Interior's and the administration's full support of the scenic-river bill but suggested that, since Interior had already formally designated the New a national scenic river, there was no need for separate designation by Congress. Kleppe suggested that the substance of the bill be changed to (1) statutorily recognize and affirm Interior's scenic-river designation and (2) provide that any license theretofore or thereafter issued by the FPC should not be permitted to inundate or adversely affect the scenic-river portion of the New. This change linked congressional action to Interior's study and designation pursuant to Governor Holshouser's application of December 12, 1974.

Taylor and Neal accepted this, and the final version of the bill, H.R. 13372, recognized as a national scenic river "that segment of the New River in North Carolina extending from its confluence with Dog Creek downstream approximately 26.5 miles to the Virginia State line."

In addition, the bill was designed to block the power project:

Any license heretofore or hereafter issued by the Federal Power Commission affecting the New River of North Carolina shall continue to be effective only for that portion of the river which is not included in the National Wild and Scenic Rivers System pursuant to section 2 of this Act and no project or undertaking so licensed shall be permitted to invade, inundate or otherwise adversely affect such river segment.

The May 6 hearing on H.R. 13372 pitted the same old foes against each other. In opposition to the bill stood American Electric Power (AEP) and the utilities industry, the AFL-CIO, the Commonwealth of Virginia, and the Ashe County Citizens Committee. In favor of the measure were members of the National Committee for the New, the State of North Carolina, and several national conservation organizations.

Joe Dowd, on behalf of AEP, testified that Blue Ridge was a truly fine project that would create two beautiful lakes in a mountain setting. He called the proposed bill a perversion of the Wild and Scenic Rivers Act because its only purpose was to block the power project. "Were we to abandon Blue Ridge tomorrow," he stated, "support for scenic river inclusion would evaporate like the morning dew. We all understand this—including the Department of the Interior which is apparently willing to let itself and the Act it administers be misused in this manner." Dowd made the additional point that the subject of the hearing was the Blue Ridge Project and whether it should be built and that the issue had already been decided in the affirmative by an expert federal agency, the FPC. He stated that the subcommittee should not, on the basis of a one-day hearing involving the unsworn testimony of nonexpert witnesses, substitute its judgment for that of the FPC, which had spent some forty thousand man-hours on the matter.

Dowd then injected a new issue into the fray. He stated that coal-fired alternatives to Blue Ridge would cost over $500 million more to build. He claimed that the license to build Blue Ridge was a property right owned by Appalachian. If Congress should pass the bill and thus deprive Appalachian of this property right, he threatened that the company would bring suit for compensation in the Court of Claims and ask for $500 million in damages from the United States. Dowd called on the subcommittee to reject the bill because "this is a staggering price to pay for the preservation of a 26.5 mile reach of river and tributary which, prior to the advent of the Blue Ridge, few people outside of its immediate vicinity had even heard of."

In support of his threat of a $500 million lawsuit against the

United States, Dowd produced a legal opinion from the Washington law firm of Covington and Burling, which concluded that Appalachian "possesses substantial arguments that legislation which in effect revokes its license . . . would constitute a taking of property under the Fifth Amendment and a breach of contract with the United States," and that Appalachian "could assert Tucker Act claims of considerable persuasiveness. . . ." The proponents of the scenic-river bill noted that the legal opinion stopped short of saying Appalachian would win its lawsuit, however.

The other principal witnesses against the bill were Congressmen William Wampler and Caldwell Butler of Virginia and Jack Curran, lobbyist for the AFL-CIO. Curran extolled the virtues of Blue Ridge: "Enhanced recreational possibilities, increased power development, flood control and economic development and jobs, both for construction workers and others in the project area, add up to a package of benefits that, in my mind, is far more than merely an effective compromise."

North Carolina Governor Holshouser and Attorney General Edmisten testified in favor of the bill, as did Steve Neal and Assistant Secretary Nathaniel Reed of the Department of the Interior. Representatives of the major conservation organizations, including the Sierra Club, Izaak Walton League, and American Rivers Conservation Council, also joined the parade of witnesses for the bill. Much of the testimony was quite technical about the need for conservation of energy. The meaning of the project in human terms was driven home, however, when several of the people of the valley testified about the strain they had been under for fourteen years and what Blue Ridge would do to their homes and lives. Hal Eaton told the subcommittee that the people of the valley had made the trip at their own expense and many had been up all night. He said, "If the people of the New River valley win, they will simply hold onto what was always rightfully theirs. If they lose, they lose their homes, churches, schools, cemeteries, farmland, and a whole way of life."

After the May 6 hearing, Appalachian's tactics became clear. It would use the threat of the $500 million suit to try to delay the bill at least and would employ some of the best lobbyists in the business to swing the necessary votes. In addition to lawyers from Covington and Burling, Jerald Segal, a lawyer with the prestigious Washington firm of Clifford and Case, was also enlisted in its cause. With Jack Curran, it had big labor on its side as well.

The scenic-river forces moved to counter this effort. Neal asked the Congressional Research Service of the Library of Congress for an opinion on the question whether the United States would have to pay compensation to Appalachian. After researching the issue, Robert Meltz, a legislative attorney, submitted a memorandum of law to the subcommittee stating that the passage of the bill would not be a compensable taking. He based his conclusion on the idea that a license is not a contract or "property" but a privilege granted by the United States and that the federal government under the law retains authority over navigable waters that precludes private ownership of the water, its flow or storage. Armed with this opinion, Roy Taylor and Steve Neal called Appalachian's threat a bluff.

The National Committee for the New River decided to hire its own full-time lobbyist for the bill. Don Kanak, a recent graduate of the University of North Carolina who had been working for the EPA in Washington, was recruited for this effort. Kanak worked closely with three members of the American Rivers Conservation Council: Bill Painter, a veteran of the 1974 fight against Blue Ridge, Fred Chapman, and Pratt Rimmel. The National Committee for the New River even put the controversy before the shareholders of AEP. At the company's annual meeting on April 29, Ham Horton, who owned a few shares in the company, took the floor and urged company officials to abandon the project voluntarily as a gesture of goodwill.

Roy Taylor, as chairman of the Subcommittee on National Parks and Recreation, was prepared to blunt any major lobbying effort against the bill. He knew he had the votes and he moved to report the bill out before the lobbying could get started. At a subcommittee meeting on May 10, the bill was quickly approved. One of the subcommittee members, Congressman Keith C. Sebelius of Kansas, told Taylor, "We've worked as a team. I know this is important to you. The baby's all yours."

Now it was up to the full House Interior Committee. Taylor was at work there too. On May 17 he moved to put the New River bill on the Interior Committee's agenda for Wednesday, May 19. In preparation for this meeting, he called on old commitments he had built up with his colleagues during his sixteen years on the committee. Interior helped too. Secretary Kleppe wrote hand-delivered letters to each of the Republican members of the committee, which stated that President Ford personally supported the bill and opposed the power project. This was too much for the opponents, who had not had time to organize. The bill was reported out by the Interior Committee by a vote of fifteen to two. Even the two who opposed it did so only after a public apology to Taylor. Jack Curran was visibly upset by the speed of the committee's action. He cornered one of labor's frequent supporters, Phillip Burton of California, who had voted for the bill, and shouted at him, "This is totally unjustified—totally!" Burton, embarrassed, quietly hustled the red-faced Curran out of the room.

In the Senate, the New River bill (S. 158) started moving also, under pressure from Helms and the lobbying effort by the National Committee for the New. Morgan, who had still not announced his decision, succeeded in postponing hearings by the Senate Interior Committee one more time, on May 13, but it was evident this would be the last delay. Senator Metcalf set the bill for two days of hearings on May 20 and 21. These hearings were essentially a replay of the House subcommittee hearing of May 6. The highlight of the session came, however, when Senator Morgan appeared to testify. He was beaten and he knew it. In a long and rambling statement before the committee, he withdrew his opposition to the bill. He reiterated his position, however, and maintained he was right in standing alone to protect landowners' rights. He said his concerns, however, had been satisfied by Carl's assurances that the state would not use floodplain zoning to implement the scenic-river management plan.

Our strategy had worked; Morgan was in the fold. Along the river, most people felt that Morgan's issue had been phoney from the beginning. They realized that the river handles its own "zoning." Walter Neaves of Crumpler said that Morgan needn't worry that the state's zoning would keep people from building on the New River's bottomlands because the river had already done that: "After the 1940 flood, the residents here built their houses farther up the hillsides, out of the floodway."

After the hearing, the Senate Interior Committee quickly approved the New River bill. On June 3, by a roll-call vote of seven to three, it was approved; a few days later it was reported favorably to the Senate. Its approval had been accomplished in an amazingly short period of time, especially considering Morgan's initial opposition.

Although the Interior committee in both the House and the Senate had reported identical New River bills, everyone on both sides realized that the real test was yet to come. The bill still had to pass the House Rules Committee. The defeat of the 1974 bill by that body was still an open sore. Would it happen again? AEP and the AFL-CIO hoped for a replay of the 1974 scenario. The National Committee for the New vowed that, this time, it would be ready.

Roy Taylor began the effort to convince the members of the Rules Committee a few days after the New River bill was reported out of the House Interior Committee. He wrote each member a personal letter, explaining that the bill was especially important to him because he was retiring from Congress. "Before I leave Congress," he wrote, "I hope I can tell my folks back home, regardless of the outcome on the New River legislation, that every member of the House has a chance to consider what action should be taken with respect to this valuable resource." Taylor also reminded each Rules Committee member that he had played a part in establishing a park or recreation area in his particular district.

Despite this effort, by the middle of June it was evident that prospects for the bill's approval by the committee were not bright. As in 1974, labor was making severe inroads among Rules Committee members. George Meany, AFL-CIO president, personally lobbied many congressmen. Of the sixteen-member committee, only six were counted as committed to the legislation.

The National Committee for the New River decided something was needed to draw attention to the issue. Ham Horton decided to hold a New River Appreciation Day in the Senate Caucus Room on June 21. Members of Congress were invited to stop in for a sample of mountain food and bluegrass music and to inspect the handicrafts made in the valley. The National Committee arranged for various speakers, including Dr. Larry Tise, head of the North Carolina Division of Archives and History. The entire North Carolina congressional delegation stopped in, and Holshouser and Edmisten also attended. The man who stole the show, however, was the National Committee's "secret weapon," former Senator Sam Ervin, who walked the halls of the Capitol signing autographs and holding forth on the beauties of the New River. He also made two speeches in favor of the scenic-river bill, saying that the power company and labor had formed an "unholy alliance" to stop the legislation and that the FPC had been guilty of "utter contempt for Congress" in licensing Blue Ridge. Ervin maintained that the almost "unlimited supplies of coal" should be used for power generation instead of the New River, "which is the handiwork of Almighty God, which was present when the morning and the evening stars sang together at the dawn of creation." He told the crowd, "There's only one New River in the world, and if it's destroyed and the country turned into mud flats, the Lord God Almighty with assistance of the AFL-CIO and all the power companies in the world cannot replace it."

But as of June 30 the New River bill did not have the votes to make it out of the Rules Committee. One vote had been gained, making a total of seven in favor; five had announced they would vote against the bill. Two of these five, Spark Matsunaga of Hawaii and Joe Moakley of Massachusetts, confessed they were really in favor of the bill but were under heavy pressure from labor lobbyists. (Matsunaga was in a fight for the nomination to run for the U.S. Senate and needed labor's endorsement.) Four members of the committee, Delwin Clawson of California, James Delaney of New York,

John Young of Texas, and Trent Lott of Mississippi, were undecided. The backers of the bill decided to hold off trying to get a vote
in the Rules Committee until after the Democratic National Convention in July. There was no use going before the committee
without the votes to win.

The National Committee for the New River by late June had
organized the elements of the last-ditch effort to get the bill out of
the Rules Committee. First, it was necessary to generate mail and
personal visits to committee members by their own constituents.
Each member of the National Committee was urged to get in touch
with friends or acquaintances in those districts. Bill Painter contacted local conservation organizations in committee members' districts. In a short while each of the congressmen on the committee
was hearing about the New River from the folks back home. Wallace Carroll also generated newspaper columns and editorials in
the home districts of committee members.

Second, national press coverage was sought and obtained. Walter
Cronkite of CBS, Harry Reasoner of ABC, and David Brinkley of
NBC were approached about featuring the issue on their evening
television news programs. All did so from time to time. Through
the efforts of Wallace Carroll, Bill Painter, and chapters of the National Committee for the New River located in other states, over
eighty newspapers across the country urged the Rules Committee
to release the bill. On June 24 the issue was featured on the program "All Things Considered" on the National Public Radio
Broadcasting System. By mid-July the New River had become a
truly national issue.

Third, a lobbying effort was undertaken headed by Kanak and
Painter, who coordinated their efforts with the North Carolina
congressional delegation. Friendly congressmen, such as Paul Simon of Illinois, were invited to go on canoe trips on the river.
Simon, after returning to Washington, became one of the most effective advocates of the bill. Contacts were also maintained with key
staff people of the Rules Committee members.

People from near and far joined in the effort. Farmers from the
New River Valley walked the halls of Congress. Earl Hamner

stepped into the fight again by writing to each Rules Committee member from his studio in Burbank, California. Jim Hunt, the lieutenant governor of North Carolina, who was running for the 1976 Democratic nomination for governor, wrote to all his fellow lieutenant governors in the home states of Rules Committee members on behalf of the bill. Governor Holshouser even proclaimed July 18, 1976, as an official "Day of Prayer for the New River."

The National Committee for the New River and the American Rivers Conservation Council, working above a fast food restaurant on Pennsylvania Avenue in Washington, turned out position papers and fact sheets on virtually every aspect of the power project for delivery to members of Congress. Ham Horton used every possible opportunity to debate Appalachian and AEP officials on the merits of Blue Ridge. Miles Bidwell, then an economics professor at Wake Forest University, provided a study which concluded on the basis of existing plants that coal-fired alternatives would be cheaper per kilowatt-hour than Blue Ridge.

An unexpected boost came to the dam opponents when Blue Ridge and the AEP were attacked in court by some small towns in Indiana and Michigan. These cities, Richmond, Indiana, and Niles, Michigan, operated municipally-owned electric companies with their own coal-fired generating facilities. Their utility systems were interconnected with a subsidiary company of the AEP system, Indiana & Michigan Power Company (I&M), and from time to time were wholesale customers of I&M. In November, 1975, I&M had discontinued the cities as wholesale customers and announced its intention to withdraw from the wholesale market. The practical effect of this, the cities felt, would be to force them to sell out to I&M and become a part of its retail system. On April 26, 1976, they filed a petition to intervene in the Blue Ridge licensing proceeding before the FPC. They asked the FPC to hold hearings on the anticompetitive effect of Blue Ridge and possible violations of the antitrust laws if the power project was to be operated only for retail customers of the AEP system.

Supporters of the New River bill also had to contend with the other major opponent, the AFL-CIO. From the first, it was obvious that the power company's clout with the Rules Committee was minimal, but labor was another matter. Fifteen of the sixteen members had received substantial contributions from labor to finance their 1974 campaigns, ranging from $3,250 for Sisk of California to $16,600 for Pepper of Florida. The New River Committee could not match this clout. All it could do was emphasize the grass-roots support for the bill and portray the opponents as "fat cats." In July each member of the Rules Committee received a gallon jug filled with the clear water of the New and a "New River Pet Rock," taken from the bed of the stream.

In addition, the proponents of the bill tried to chip away at the solid front of labor opposition in advance of the Rules Committee vote. Donald Frey, a professor of labor economics at Wake Forest University, wrote to George Meany, stating that labor's opposition was hurting the embryonic labor-union movement in North Carolina. Wilbur Hobby, head of the North Carolina AFL-CIO, quietly worked behind the scenes to change labor's mind. About thirty-five residents of the New River Valley picketed AFL-CIO headquarters in Washington on July 28. Some individual unions were persuaded to announce that they favored the bill. The Amalgamated Meat Cutters International, the Amalgamated Clothing Workers of America, and the United Mine Workers all wrote Rules Committee members in favor of the bill. The United Mine Workers' endorsement was easy to get, since its members mined coal, the principal alternative to Blue Ridge. Yet Meany refused to budge.

The New River bill was by late July one of the most heavily lobbied measures of the 95th Congress. Every member of the Rules Committee had a stack of stop-the-dam mail and was questioned daily by reporters and lobbyists from both sides as to how he stood. Yet there was still no break in the committee. Only seven votes could be counted on. On July 22 normally mild-mannered Steve Neal lost his cool and angrily attacked the holdouts on the Rules Committee. "It's an outrageous distortion of the democratic process," Neal declared. "They're being led by the nose." This only succeeded in antagonizing Ray Madden, the chairman of the committee and a supporter of the bill. Madden defended the right of the members of his committee to vote their conscience. Neal was working like a man obsessed; he hounded the undecided members of the committee. Joe Moakley of Massachusetts was feeling the pressure. "I swear, it's like turning around and looking in a mirror. Every time I look, he's there."

Even though the proponents of the bill did not have the votes, time was growing short, and they could not afford to wait forever before facing the Rules Committee showdown. Congress would recess for two weeks in August for the Republican National Convention. After they returned, the members would be anxious to adjourn again to go out on the hustings and campaign. On July 30 Chairman Madden scheduled the crucial Rules Committee vote for Wednesday, August 4.

In the first few days of August, both sides got in their final shots. The proponents of the bill were not optimistic, and Neal made plans to circulate a petition for discharge signed by a majority of the total membership of the House in order to get the measure to the floor of the House in case the Rules Committee voted it down. The national news media took note of the coming test for the bill. The network news programs featured the New River issue, and many newspapers all over the country published editorials. The *Philadelphia Inquirer*, for example, ran the following piece on August 2:

Near the mountainous border of Virginia and North Carolina, that section of the New is a glory of free-flowing water, a treasure of prehistoric archeological remains, a fragile haven of rare wildlife, a stable and proud community of countryfolk whose forebears tended the land before the American Revolution.

It is also a temptation for the American Electric Power Company, which is trying to obliterate it under two dammed-up lakes for a peakload, pumped-storage generating project which by the company's own submission could be half silted in within 20 years and virtually unproductive within 50.

The immensely influential and free-spending power lobby has been joined in its war against the New River by an even more powerful one— the AFL-CIO and its construction unions, which want jobs, dues and the reconfirmation of their power which winning the fight would yield.

In 1974, majorities in both the Senate and the House declared by votes of 49 to 19 and 196 to 181 that they wanted the New River saved from the dammers. But the Rules Committee won that round, blocking the salvation effort. The Ford Administration, through Interior Secretary Thomas S. Kleppe, has done all possible to save the river.

Now once again, the Rules Committee, or a substantial number of its 16 members who are responsive to heavy lobbying pressure, holds the fate of the New River in its hands. . . .

If enough constituents made clear to each congressman their demand for favorable action by the Rules Committee, or if that fails for the swift execution of a discharge petition to put the debate before the whole House, the New River could be saved, and an important blow would have been struck against minority rule for special interests' sake.

The atmosphere was tense when the Rules Committee began its debate on the bill on August 4. It was again essentially the Virginians against the North Carolinians as the now-familiar arguments for and against the bill were repeated by both sides. Former Congressman Wilmer Mizell, who was running for Neal's seat, was allowed to speak, and he begged the committee members to report the bill.*

The Rules Committee hearing climaxed after four hours into an angry debate over the remark Neal had made about the reluctance of the committee to act being a distortion of the democratic process. Republican members of the committee accused Neal of impugning the integrity of the members and called his comment a "shame and a disgrace." Neal at first defended his statement, but later, just before the vote, publicly apologized to the committee. Right up until the roll was called, Neal and Taylor were sure of only seven of the sixteen votes.

In the end the tally stood at ten in favor and six against—the

---

* Three hundred miles away, in the New River Valley, the atmosphere was tense also. The people knew that their fate and the outcome of a fourteen-year fight hinged on one committee vote. On his farm on the New River, Steve Douglas was working in his pole-bean patch but could not get the committee vote out of his mind. Suddenly, he knelt down on the soil, stretched his arms to heaven, and prayed for a favorable vote.

scenic-river bill was reported to the House. The New River had won! Joining the seven who had announced their support were two Republicans, Clawson of California and Latta of Ohio, and one Democrat, Moakley of Massachusetts. Moakley said that he had always leaned toward approval and was convinced by the debate. The two Republicans had come over because of the last minute intervention of John Rhodes, the Republican leader in the House, who had passed the word that President Ford and the administration wanted the bill. "I don't think," Rhodes had said, "the Rules Committee ought to block this bill from the full House, a bill with a national constituency and strong regional appeal." The Republicans, however, played a little politics and tried to give Mizell credit for convincing Latta and Clawson, to aid him in his election fight against Neal.*

The opponents of the dams were jubilant but cautious. A seemingly impossible hurdle had been overcome, but it was still necessary to get full House and Senate approval. In the past, something had always happened at the last minute to turn victory into defeat. Taylor and Neal moved to get an immediate House vote on the New River bill. Speaker Carl Albert was cooperative and set the measure down for August 9.

The debate in the House stretched over two days, and it was evident that labor and the power companies had not given up. James Quillen of Tennessee offered a ridiculous amendment to allow the construction of Blue Ridge but to designate the entire 236-mile length of the New River as a scenic river. "I feel that this would be a great thing for nature lovers, for conservationists and for ecologists in protecting the river which they say the first stars of heaven shone down upon," Quillen stated sarcastically. This amendment was voted down by voice vote.

The debate on the merits of the bill showed the widespread support the scenic-river bill enjoyed. Only the Virginia congressmen and Quillen spoke against it, stating that the measure was a threat to the independence of the FPC. In contrast, congressmen from various sections of the country joined the North Carolinians in

* Mizell was defeated.

supporting the bill. Seiberling of Ohio emphasized the mistakes that had been made in the past:

We have a powerplant near Akron, Ohio, on a beautiful gorge, which is served by a dam that covers up what was once a scenic waterfall on the Cuyahoga River in the town of Cuyahoga Falls. Some years ago the president of the power company told me that the company now realized that they should not have built that dam and powerplant there, and that by doing so it had deteriorated the scenic and recreational assets of the area. He acknowledged that the dam and powerplant could have been built somewhere else and today would have been.

Simon of Illinois talked about the trip he had taken down the river:

I got one of these circulars through the mail about this project, and just out of curiosity I signed up for a trip down the New River and took my family along. I will say to the members that if any of them would have the chance to take that trip on the New River through North Carolina and Virginia, this area that is to be destroyed, there would be few if any votes against this bill.

Ullman of Oregon mentioned the arrogance of the FPC when, in 1974, it licensed Blue Ridge even though the scenic-river legislation was pending in Congress at the time:

We have heard it said that this legislation has the effect of reversing the action taken by the Federal Power Commission. Never before, we have been told, has the Congress asserted itself in matters of this kind. Indeed, it may be true that this is a unique situation.

This may be a unique situation not because of congressional intervention, but because the action of the FPC in this case amounted to an ultimatum to the Congress. In effect the agency gave Congress six months to act—the last 6 months of the 93d Congress. It took this action in spite of the requests of the chairmen of the Committees on Interior and Insular Affairs asking that the decision on the license be deferred until Congress had a reasonable opportunity to act. Certainly, it is not a common practice for an agency to require the Congress to reverse its action. It is even less common for an agency to tell the Congress that it has only 6 months to act—particularly in the final weeks of a session when so many issues are confronting Congress.

Preyer of North Carolina stated eloquently what the issue meant to the people of his state:

North Carolina is a State that came late to the industrial revolution. Many of those who measure such things would say that for much of our history we have been a poor State.

Yet we have been the beneficiaries of two great sources of wealth: The human resources of our people and the natural resources of our land and water.

This bill before you seek[s] to preserve a great natural resource and, in doing that, it serves also the preservation of that human resource as well.

It is a part of the character of our people that we have great respect for the land and for the water that nourishes that land. In the New River all of this—land, water, and people—is met in harmony that is lost for much of America. We ask you not to be a party to destroying that.

Others can better here argue the facts. You know of the antiquity of the river. Know also that a people who do not respect this kind of unique reservoir of time and history are a people who are destined to be not as great as they could be, not as proud as they ought to be, not as strong as those who have preserved the land and the river before them hoped they would be.

We are proud of the New River in North Carolina. It gives for those of us in our State, as I am certain it does for those in the other States through which it runs, a link with our past and satisfaction in the simple joys of living that is lost when such resources are destroyed or distorted.

What kind of price can you put on such a treasure? How can one measure it against the relative value of better energy or cheaper kilowatts?

That is a question that each man must answer for himself in his own life as he chooses what will be important to him, what value he will seek to preserve.

Clearly the people of our State have overwhelmingly said they choose to preserve the New River as it is.

It is a unity that is as real as it is rare. The effort to save the river brings together conservative Republicans and moderate Democrats. It is supported in my State—and I believe this fact needs special emphasis—

it is supported in my State by organized labor. Few causes have so welded together people of divergent views and stations in life.

When the vote came in the House on August 10, the New River bill passed with surprising ease, 311 to 73. Even five members of the House from Virginia supported it, breaking away from the die-hard members of that state's congressional delegation. Many commentators predicted that the bill's passage, if concurred in by the Senate, would have far-reaching implications: the era of unquestioned acceptance of large dams proposed by federal agencies might be over.

Now only the Senate remained to be convinced. The supporters of the New River were quietly optimistic about the bill's chances for success. The makeup of that body had changed little since 1974, when it had overwhelmingly approved the previous New River bill.

The power-company forces decided to make one last-ditch effort to block the legislation, however. Senator William Scott of Virginia, leader of the pro-dam group, successfully obtained a week's delay in Senate consideration, which had been scheduled for August 23. Joe Dowd of AEP renewed his threat to sue the United States for $500 million in damages if the New River bill was enacted. A friendly newspaper was enlisted to try a final media blitz of pro-dam publicity. The *Chicago Tribune* sent one of its reporters, Bob Wiedrich, to spend a week in Ashe County, and every day, Monday through Friday of the week of August 23, the *Chicago Tribune* ran one of his columns. Wiedrich conducted interviews with the proponents of Blue Ridge:

"For the past 30 years, people have been migrating from the area, leaving behind only the very poor and the very old," said Dr. James E. Rhodes, an optometrist and an Ashe County commissioner. . . .

"In 1975," Dr. Rhodes continued, "we averaged 25 to 30 per cent unemployment in Ashe County. In one week, the figure hit 48 per cent. That's damned rough on people.

"This project will generate a construction payroll of more than $200 million during the five to six years it takes to build it. It will produce between 1,500 and 2,000 jobs.

"And the 40,000-acre mountain lake the Blue Ridge power project would create could furnish one of the finest recreational havens in the Eastern United States."

Wiedrich then quoted J. C. Jenkins, the Jefferson tire dealer who had frequently condemned the opponents of the dam:

"It is the same kind of controversy that could occur many times more in the United States as the environmentalists and power companies go head-to-head. Somebody is going to have to make a stand against this sort of thing now.

"This county needs the economic benefits and employment to be derived from the project.

"All we've got to sell here is good climate, good air, and good water. And that project should give us the finest mountain lake in the world.

"The last thing we need is a scenic river designation because that places restrictions on the use property owners along the river can put to their land. It's restricting what they can do without compensation."

But it was a case of too little, too late. The National Committee for the New was at work too, and a final avalanche of letters and telegrams poured into the Senate. Helms and Morgan personally contacted their fellow senators to enlist them to vote for the measure. Pickets from the New River area marched around Capitol Hill, handing out brochures praising the river.

In addition, as the Senate moved to a vote, George Meany was having second thoughts about his support for the power project. He was having trouble holding labor together, and he realized that he was about to go down to defeat. Accordingly, in late August, he decided to withdraw AFL-CIO opposition to the bill. In a personal letter to Joe Matthews on August 19, Meany explained his new position:

The special nature of the New River and the questionable need for, and utility of, the proposed dam make our past support for construction of this dam open to the serious questions which you and many others have raised. I think our support for it has to be seen against the background of the Nation's need for new energy sources and the broad scale efforts of some environmental groups to thwart every power project of whatever type. The combination of those two factors has led the

AFL-CIO and a number of other organizations to support the development of new energy sources and to defend the established procedures
for dealing with challenges to this type of project. . . .

In the case of the New River Project, we have followed a line of
conduct which logically follows from these considerations. It was not
a capricious decision and it was not dictated solely by the construction
jobs involved, as some critics have alleged. The self-evident power needs
of the nation and the long-term benefits to the economy that would
result were our prime consideration.

It is now obvious to me that there were other factors and developments that we did not fully explore before testifying this year. The real
impact of the dam on the people who live in the valley and the destruction of the natural beauty of the area, were really only brought home
to me very forcefully during the past few weeks by strong representations made by some of our trade unionists in North Carolina, by letters
such as yours, and by representations made by a group of residents and
others who visited in our office in Washington.

I am sure you will understand that in saying that, I am not trying to
avoid responsibility for the positions which we took, but rather only to
explain them.

Labor's change of heart was a final indication that victory was
at hand. When the Senate took up consideration of the bill on August 30, two Virginia senators, William Scott and Harry Byrd, Jr.,
tried valiantly to muster support for the power project. Both Helms
and Morgan spoke eloquently in favor of the scenic river. Governor
Holshouser sat in the gallery watching the debate. Almost all the
other senators who joined the debate spoke against the dams. One
of the highlights of the proceedings occurred when Senator Barry
Goldwater got up to express his opinion:

I rise to announce my opposition to this dam. . . . Of all the votes
I have cast in the 20-odd years I have been in this body, if there is one
that stands out above all others that I would change if I had the chance
it was a vote I cast to construct Glen Canyon Dam on the Colorado.

. . . While Glen Canyon Dam has created the most beautiful lake in
the world and has brought millions and millions of dollars into my State
and the State of Utah, nevertheless, I think of that river as it was when
I was a boy and that is the way I would like to see it again.

The final vote was sixty-nine to sixteen in favor of the bill to affirm the scenic-river designation of the New. President Ford immediately told Governor Holshouser he would sign it. Congress, after so many long years, had acted to save the river. The deadlock between Interior and the FPC had been broken in favor of the New River. Blue Ridge was dead.

The sun shone brightly on the morning of September 11, 1976, when fifty invited supporters of the New River bill filed into the White House rose garden to attend the signing by President Ford. Before penning his signature, Ford said, "This majestic and beautiful river and the land surrounding it have been preserved for future generations. I hope the New River will flow free and clear for another 100 million years." After the ceremony, Governor Holshouser, in a lighter moment, thanked the President for his support and presented him with a charter certificate naming him "Admiral of the New River Navy."

To the people of the New River Valley and the many thousands of their allies who had worked to preserve the river, it was a day of profound joy. There was no cheering, no dancing in the streets—only quiet and deep satisfaction and thankfulness. The people who were most involved didn't say much about their victory; words were not adequate to express the depth of their feeling. When they did have something to say, it was not about the victory over the power company but only about the way of life in the valley and how it would remain just as it had been.

President Ford, after signing the bill protecting the New River, commented: "When a decision has to be made between energy production and environmental protection . . . you must ask what is the will of the people involved. . . . It is clear in this case the people wanted the New River like it is."

Later that day, on the "NBC Evening News," a national television audience saw Elizabeth McCommon, beaming and beautiful, standing high above New River, the valley green and lush in the background. In a voice full and rich she sang out the celebration of the people:

How rich, how green is my valley;
That precious place my New River home.
I can hear the torrent roar,
I can see the high hawk soar
From that spot of land my father gave to me.

Epilogue

THE STORY OF THE NEW RIVER SHOWS IN CON-
crete terms the folly of a policy of unlimited development of all
available resources. This policy, had it been carried out, would
have created a power project that consumed four units of electricity
for every three it produced and would have destroyed a deeply-
rooted, traditional mountain culture. Yet, because of our national
commitment to "cheap and abundant" energy, this was almost allowed to happen by persons who gave no thought to conservation,
persons with the attitude that there are essentially no limits to
either our use of or our ability to produce energy.

The successful fight to preserve this small but beautiful river valley in Appalachia was accomplished by a unique coalition of diverse groups and interests. The people of the New River Valley
were struggling for their homes and farms. The environmentalists
wanted to save a free-flowing mountain stream. The State of North
Carolina opposed the appropriation of state resources to produce
power that would be exported to other areas of the country. Liberals
cited the environmental and aesthetic degradation of the project.
Conservatives were concerned about losing traditions and the way
of life of the valley's people. None of these interests would have
prevailed alone; together they were a formidable force.

It is remarkable that the effort to save the New River was mounted
successfully at a time when the need for new supplies of energy
was unquestioned. Long lines at gasoline pumps and the Arab oil
embargo had brought home the vulnerability of the country to
energy shortages. Other energy development projects such as the
Alaskan pipeline and the opening of the submerged lands of the
outer continental shelf to oil and gas exploration were approved by
Congress and the federal agencies. Yet the New River was spared.

The fundamental reason for this is that the plight of the people
and the river struck a populist chord in the mind of the country
that resounded more loudly than the concern with energy shortages.
This was not a wilderness being threatened, but a simple, traditional human culture of small farms utilizing—but living in harmony

with—the natural world. The agents of destruction were large corporate and labor interests and their allies in the federal government. To many people this conflict represented a struggle to preserve something of their own roots in the American experience and the relationship with the land. The New River controversy proved that these profound values still exist in America despite energy shortages and economic concerns.

Even so, the struggle for the New River was an epic political and legal battle involving the major institutions of our society. The entire congressional delegation as well as the government of the State of North Carolina became actively involved. The Department of the Interior and the administration of President Ford entered the fray on the side of the river. Yet these powerful political forces were only barely able to defeat the will of the FPC, the labor unions, and the largest private utility company in the nation. The courts played a key role in the matter, although they refused the opportunity to take forceful action. In the end, the Congress was the arena in which the impasse was broken.

Although the New River itself was always the center of the controversy, its own merits were obscured by the technicalities of the legal procedures and the institutional manipulations. The right of the river itself to exist was rarely, if ever, addressed. There is no mechanism in the law to confront directly the issue of whether a natural feature that has existed for 100 million years should be fundamentally altered by man. We have no criteria for such a determination. In the end, however, the river was vindicated and allowed to remain free because it was perceived as important to the needs, dreams, aspirations, and values of many thousands of people.

It may be that few environmental controversies in the future will have the drama or appeal of the New River. But there is much that this battle can teach us. We know that we can survive and prosper as a society without developing all our available resources. It is both economically and environmentally sound to utilize presently available energy systems more efficiently rather than to depend totally upon the development of new supplies. We can make our industrial processes, automobiles, buildings, and consumer goods more energy efficient. Our economic well-being will be damaged, not enhanced, by the compulsion to invest in ever more costly energy supply projects.

Another lesson of the struggle to preserve the New is that environmentalist values alone are not enough to carry the day in the political arena. Concern with wildlife and aesthetics are too narrow and elitist to appeal to a broad segment of society. Places of natural beauty and historical importance like the New River will be preserved only if a broad cross section of the population is convinced that they are part of our heritage as a people. The management of our resources is inseparable from the context of the American cultural tradition.

Bibliography

The material in this book has come primarily from documents and letters, newspaper articles, interviews, and my own experience. Selected sources are included in the bibliography.

1/THE RIVER AND THE VALLEY: GEOLOGY AND PREHISTORY

Ayers, Harvard. "An Appraisal of the Archeological Resources of the Blue Ridge Project." Duplicated. Washington: River Basin Surveys, Smithsonian Institution, 1965.

Clay, James W.; Orr, Douglas M.; and Stuart, Alfred W., eds. *North Carolina Atlas*. Chapel Hill: University of North Carolina Press, 1975.

Holland, C. G. "The Revised Blue Ridge Project: An Archeological Survey and Summary." Duplicated. 1969.

Janssen, Raymond E. "The Teays River, Ancient Precursor of the East." *Scientific Monthly*, LXXVII (December, 1953), 306–14.

Rights, Douglas L. *The American Indian in North Carolina*. Winston-Salem: John F. Blair, Publisher, 1957.

Willey, Gordon R. *An Introduction to American Archeology: North and Middle America*. Vol. 1. Englewood Cliffs, N.J.: Prentice-Hall, 1966.

2/THE NEW RIVER IN THE OLD WEST: CONFLICT AND SETTLEMENT

Arthur, John P. *Western North Carolina: A History*. Asheville, N.C.: Edward Buncombe Chapter of the DAR, 1914.

Caruso, John A. *The Appalachian Frontier: America's First Surge Westward*. Indianapolis: Bobbs-Merrill Co., 1959.

Cox, Aras B. *Footprints on the Sands of Time: A History of Southwestern Virginia and North-western North Carolina*. Sparta, N.C.: Star Publishing Co., 1900.

Draper, Lyman C. *King's Mountain and Its Heroes*. Baltimore: Genealogical Publishing Co., 1967.

Driver, Carl S. *John Sevier: Pioneer of the Old Southwest*. Chapel Hill: University of North Carolina Press, 1932.

Heavener, Ulysses. *German New River Settlement*. Baltimore: Genealogical Publishing Co., 1976.

Henderson, Archibald. *The Conquest of the Old Southwest*. New York: Century Co., 1920.

Henderson, Archibald. *The Star of Empire: Phases of the Westward Movement in the Old Southwest*. Durham, N.C.: Seeman Printery, 1919.

Lefler, Hugh, and Newsome, Albert. *North Carolina: The History of a Southern State*. 3rd edition. Chapel Hill: University of North Carolina Press, 1973.

Messick, Hank. *Kings Mountain: The Epic of the Blue Ridge "Mountain Men" in the American Revolution*. Boston: Little, Brown and Co., 1976.

Newsome, A. R. "Twelve North Carolina Counties in 1810–1811." *North Carolina Historical Review*, V, No. 4 (October, 1928), 413–46.

Ramsey, R. W. *Carolina Cradle: Settlement of the Northwest Carolina Frontier, 1747–1762*. Chapel Hill: University of North Carolina Press, 1964.

Summers, L. P. *Annals of Southwest Virginia*. Abingdon, Virginia: Lewis Summers, 1929.

3/THE LOST PROVINCES

Bake, William A. *The Blue Ridge*. New York: Viking, 1977.

Barrett, John G. *The Civil War in North Carolina*. Chapel Hill: University of North Carolina Press, 1963.

Fletcher, Arthur. *Ashe County: A History*. Jefferson, N.C.: Ashe County Research Association, 1963.

Tise, Larry. "Treasures of the New River Valley." *The State*, XLIX, No. 4 (September, 1976), 18.

U.S., Dept. of the Interior, Bureau of Outdoor Recreation. *Final Environmental Statement: Proposed New River Gorge National Wild and Scenic River in West Virginia*. 23 July 1976.

U.S., Dept. of the Interior, Bureau of Outdoor Recreation. *Final Environmental Statement: Proposed South Fork New River National Wild and Scenic River, North Carolina*. March, 1976.

Van Noppen, Ina W. and John J. *Western North Carolina since the Civil War*. Boone, N.C.: Appalachian Consortium Press, 1973.

4/"PROGRESS" COMES TO THE NEW RIVER: PHASE ONE OF THE BLUE RIDGE PROJECT

Kenworthy, E. W. "Pollution Dilution Issue in Blue Ridge Power Plan." *New York Times*, November 7, 1971, Sec. 1, p. 58.

U.S., Federal Power Commission. *Order Reopening the Proceeding and Requiring Further Procedures to Implement the National Environmental Policy Act*, 2 November 1972.

U.S., Federal Power Commission. *Presiding Examiner's Initial Decision Upon Application for Blue Ridge Hydroelectric and Pumped Storage License*, 1 October 1969.

U.S., Federal Power Commission. *Presiding Examiner's Supplemental Initial Decision Upon Remand—Blue Ridge Project License Application*, 21 June 1971.

5/ANOTHER GREEN LIGHT FOR BLUE RIDGE

U.S., Federal Power Commission. *Additional Supplemental Initial Decision—Blue Ridge Project License Application*. 23 January 1974.

U.S., Federal Power Commission. *Opinion and Order Granting a License for the Modified Blue Ridge Project*. Opinion No. 698, 14 June 1974.

6/THE SCENIC-RIVER STRATEGY

U.S., Congress, Senate, *Congressional Record*. 93d Cong., 2d sess., 28 May 1974, pp. 9055–73.

U.S., Congress, House. *Congressional Record*. 93d Cong., 2d sess., 18 December 1974, pp. 12261–68, 12290–91.

U.S., Congress, House, Interior Committee. *Amending the Wild and Scenic Rivers Act of 1968 by Designating a Segment of the New River as a Potential Component of the National Wild and Scenic Rivers System*. 93d Cong., 2d sess., 3 October 1974, H. Report 93–1419.

7/COUNTERATTACK IN COURT

State of North Carolina v. *Federal Power Commission*, 393 F. Supp. 1116 (M.D.N.C. 1975).

8/THE PEOPLE UNITE

Dillon, Tom. "Celebrating the River." *Winston-Salem Journal*, July 27, 1975, Sec. A, p. 1.

9/FALSE HOPES AND A NEW DEFEAT

Poole, Bob. "New's Archeological Resources Were Not Protected." *Winston-Salem Journal*, August 31, 1975, Sec. A, p. 1.

State of North Carolina v. *Federal Power Commission*, 533 F. 2d 702 (D.C. Cir. 1976), judgment vacated 97 S. Ct. 250 (1976).

U.S., Congress, Senate, Committee on Interior and Insular Affairs. *Designating a Segment of the New River, North Carolina, as a Component of the National Wild and Scenic River System*. 94th Cong., 2d sess., 16 June 1976, S. Report 94–952.

10/THE NEW RIVER LIKE IT IS

Collins, Rip. "Saga of the Undammed." *Wildlife in North Carolina*, XLI, No. 2 (February, 1977), 14–16.

Painter, Bill. "Debunking Madison Avenue." *Environmental Action*, February 28, 1976, p. 8.

U.S., Congress, House. *Congressional Record*. 94th Cong., 2d sess., 9 August 1976, pp. 8507–29.

U.S., Congress, House. *Congressional Record*. 94th Cong. 2d sess., 10 August 1976, pp. 8589–8607.

U.S., Congress, Senate. *Congressional Record*. 94th Cong., 2d sess., 30 August 1976, pp. 14929–49.

Index